Lecture Notes in Networks and Systems 1933

The series "Lecture Notes in Networks and Systems" publishes the latest developments in Networks and Systems—quickly, informally and with high quality. Original research reported in proceedings and post-proceedings represents the core of LNNS.

Volumes published in LNNS embrace all aspects and subfields of, as well as new challenges in, Networks and Systems.

The series contains proceedings and edited volumes in systems and networks, spanning the areas of Cyber-Physical Systems, Autonomous Systems, Sensor Networks, Control Systems, Energy Systems, Automotive Systems, Biological Systems, Vehicular Networking and Connected Vehicles, Aerospace Systems, Automation, Manufacturing, Smart Grids, Nonlinear Systems, Power Systems, Robotics, Social Systems, Economic Systems and other. Of particular value to both the contributors and the readership are the short publication timeframe and the world-wide distribution and exposure which enable both a wide and rapid dissemination of research output.

The series covers the theory, applications, and perspectives on the state of the art and future developments relevant to systems and networks, decision making, control, complex processes and related areas, as embedded in the fields of interdisciplinary and applied sciences, engineering, computer science, physics, economics, social, and life sciences, as well as the paradigms and methodologies behind them.

Indexed by SCOPUS, EI Compendex, INSPEC, WTI Frankfurt eG, zbMATH, SCImago.

All books published in the series are submitted for consideration in Web of Science.

For proposals from Asia please contact Aninda Bose (aninda.bose@springer.com).

Ariel Santos Fuentefria ·
Daniel Alfonso Robaina · Joel Morales Salas ·
Eduardo García Noa
Editors

Advances in Energy Efficiency, Renewable Energy and Environment

Selected Contributions to the II International Conference on Renewable Energy Sources, Energy Efficiency and the Environment (JIFREMA 2025), June 4–6th, 2025, Havana, Cuba

Springer

Editors
Ariel Santos Fuentefria
Havana, Cuba

Daniel Alfonso Robaina
Havana, Cuba

Joel Morales Salas
Havana, Cuba

Eduardo García Noa
Havana, Cuba

ISSN 2367-3370 ISSN 2367-3389 (electronic)
Lecture Notes in Networks and Systems
ISBN 978-3-032-23027-0 ISBN 978-3-032-23028-7 (eBook)
https://doi.org/10.1007/978-3-032-23028-7

This Springer imprint is published by the registered company Springer Nature Switzerland AG
The registered company address is: Gewerbestrasse 11, 6330 Cham, Switzerland

Contents

Biomethane and Biohydrogen from Vinasse: Simulation and Technical-Environmental Analysis

Javier Horta García[1(✉)], Arletis Cruz Llerena[1], Osney Pérez Ones[2], and Lourdes Zumalacárregui de Cárdenas[2]

[1] Department of Bioengineering, Direction of Biotechnology, Cuban Research Institute of Sugarcane Derivatives (ICIDCA), Vía Banca 804 y Carretera Central. San Miguel del Padrón, La Habana, Cuba
hortajavier38@gmail.com

[2] Process Analysis Research Group, Chemical Engineering Faculty, Universidad Tecnológica de La Habana "José Antonio Echeverría" (CUJAE), Calle 114 No. 11901 E/ Ciclovía y Rotonda, Marianao, La Habana, Cuba

Abstract. Vinasse in one of the industrial effluents with the highest pollutant load, and its revalorization into biofuels is one of the pillars of sustainability in the sugarcane byproducts industry. In the present work, the simulation of the production processes of biomethane and biohydrogen from distillery vinasse is carried out. The Aspen Hysys v10.0 simulator was used to develop the simulation models. The NRTL property package was selected for the liquid phase and Peng-Robinson for the vapor phase in the biomethane simulation model, while in the biohydrogen simulation model, NRTL-Ideal was used for the anaerobic digestion stage and Peng-Robinson for the hydrogen production stage. The obtained simulation models were validated with data reported in the literature, achieving an error of less than 5% in both cases. A simplified life cycle assessment was conducted, yielding emission factors of 0.027 kg CO_2eq/MJ for biomethane and −0.102 kg CO_2eq/MJ for biohydrogen. These values were compared with those of fossil fuels, confirming the climate advantage of both biofuels in reducing greenhouse gas emissions.

Keywords: Vinasse · biomethane · biohydrogen · simulation · biofuels

1 Introduction

Vinasse is a liquid byproduct generated during ethanol production from sugarcane, considered one of the most polluting streams in this industry. Its high organic load, low carbon-to-nitrogen ratio, and presence of toxic compounds such as phenols, heavy metals, and sulfates cause severe environmental damage if not properly treated. The release of untreated vinasse contaminates soil and water, increases chemical oxygen demand (COD), reduces photosynthesis in aquatic ecosystems, and emits greenhouse gases. In addition, its acidic pH and high solids concentration are harmful to biodiversity

A. Santos Fuentefria et al. (Eds.): JIFREEEMA 2025, LNNS 1933, pp. 1–11, 2026.
https://doi.org/10.1007/978-3-032-23028-7_1

and human health. However, vinasse has great potential for biogas production through anaerobic digestion, a process that not only reduces its pollutant load but also generates energy. Converting vinasse into biogas offers economic and environmental benefits, such as replacing fossil fuels, producing biofertilizers, and reducing emissions, thus transforming an environmental pollutant into a sustainable energy resource [1].

Biomethane is obtained by purifying biogas. The process removes carbon dioxide and other compounds, achieving a natural gas-like composition. Transportation, electricity generation, and domestic consumption are among its applications. High-pressure water scrubbing is a leading technology for biogas purification, thanks to its sustainability and efficiency. This method relies on water's ability to selectively absorb carbon dioxide under high-pressure conditions, leveraging the difference in solubility between methane and carbon dioxide. This technology not only achieves methane purity levels above 98% but is also environmentally friendly, as it does not require chemicals and operates at relatively low costs [2].

Biohydrogen, derived from biogas, represents a sustainable and promising solution in the energy transition toward renewable sources. Among the available methods for producing hydrogen from biogas, steam reforming is the most widely used due to its efficiency and technological maturity. This process involves the reaction of methane present in biogas with steam at high temperatures and in the presence of a catalyst, typically nickel, to produce hydrogen and carbon dioxide. The main reaction is endothermic, meaning it requires significant thermal energy input to proceed. A key advantage of steam reforming is its ability to utilize biogas methane, converting it into a high-purity hydrogen source that can be used in applications such as fuel cells or industrial processes. However, this method also generates carbon dioxide as a byproduct, posing environmental challenges that can be mitigated through complementary technologies such as carbon capture and storage (CCS). In the current context of decarbonization and emission reduction, steam reforming applied to biogas provides an efficient pathway for producing renewable hydrogen while utilizing organic waste and contributing to the circular economy [3, 4].

Simulation is a powerful and versatile tool for evaluating the potential of biofuel production, allowing the exploration of different scenarios and process optimization without the costs and risks associated with large-scale physical experimentation. Through computational models, key variables such as feedstock selection, reactor design, operating conditions, and the integration of different technologies can be analyzed, identifying the most efficient and sustainable configurations for producing specific biofuels. Additionally, simulation facilitates the assessment of the economic and environmental impact of processes, contributing to informed decision-making and the development of strategies for the successful implementation of biomass-based energy.

In this work, simulation models in Aspen Hysys v10.0 are developed and validated for the production of biomethane via high-pressure water scrubbing and biohydrogen via steam reforming, using distillery vinasse. These models can be used to assess the potential of vinasse in biofuel production. A simplified life cycle analysis is also conducted for both processes to evaluate their ability to reduce greenhouse gas emissions.

2 Materials and Methods

2.1 Process Description

Both production processes begin by cooling the vinasse from the distillation column to 55 °C to ensure that the anaerobic digester operates in thermophilic conditions, as this temperature promotes optimal microbial growth and methane productivity under the selected conditions. The cooled vinasse is then fed into the biodigester, where biogas is produced. This biogas undergoes a desulfurization process, resulting in treated vinasse and purified biogas.

Biomethane Production via High-Pressure Water Scrubbing. The desulfurized biogas is compressed to 1 MPa and introduced at the bottom of the absorption column, where it is scrubbed with water to remove CO_2 [5]. The resulting biomethane passes through a dryer to eliminate residual water before being compressed to 20 MPa for distribution.

The CO_2-rich water is sent to a separator where the pressure is reduced from 1 MPa to 300 kPa to minimize methane losses. The gas exiting the separator is mixed with the raw biogas and recirculated to the compressors. The aqueous solution is passed to a desorbing column where it contacts air to remove the remaining CO_2 and H_2S. The water is then recirculated back to the absorption column [5].

Biohydrogen Production via Steam Reforming. The desulfurized biogas is compressed to 2.5 MPa and heated to 700 °C before entering the reformer, together with steam at the same pressure and temperature. The methane steam reforming reaction occurs at a steam-to-methane ratio of 4 [6].

The reformer effluent is cooled to 350 °C before entering a water-gas shift reactor. The products from this reactor are further cooled to 210 °C before entering a second low-temperature shift reactor. The output from the low-temperature reactor undergoes pressure swing adsorption (PSA), where hydrogen is separated from the residual gas. One-third of the residual gas is recirculated to the reformer inlet to enhance process efficiency, while the remainder can be used as fuel. The purified hydrogen is compressed to 60 MPa and cooled to 20 °C for distribution [7].

2.2 Development of Simulation Models

The simulation of biomethane and biohydrogen production processes was performed using Aspen Hysys v10.0. The general procedure for the simulation included the following steps: configuring the simulation (units, reporting options, etc.), selecting or creating the components involved, selecting the method for calculating the physical properties of the components, developing the flow diagram (modules and streams, typically done in the worksheet), specifying the stream data, and defining the parameters of the modules used.

For the biomethane production simulation, the NRTL property package was used for liquid-phase equilibrium (water with organic compounds) and Peng-Robinson for the vapor phase. In the case of biohydrogen production, the same property package (NRTL) was applied only in the anaerobic digestion stage, while the Peng-Robinson model was used for the hydrogen production stage due to the high pressures (2.5 MPa) and absence of a liquid phase [8, 9].

Both simulation models incorporated the anaerobic digestion model proposed by Cruz [10] for biogas production and electricity generation, with the electricity generation section removed. A vinasse mass flow rate of 17,860 kg/h was supplied, as obtained by Cruz et al. [11] for a typical distillery. Tables 1 and 2 show the mass composition of the vinasse used and the chemical reactions inserted into the anaerobic digester, respectively.

Table 1. Mass composition of the vinasse used

Component	Mass Composition (%)
Ethanol	0.41
Water	93.95
Acetic Acid	0.03
Dextrose	4.5
Sucrose	0.36
Glycerol	0.08
CaO*	0.13
C3H6NO2S*	0.09
Impurities*	0.03
K2O*	0.31
Na2O*	0.07
MgO*	0.01
C3H6O2	0.01
C4H8O2	0.02

* Hypothetical components

Table 2. Chemical reactions inserted into the anaerobic digester

Reactions	Limiting Component	Fractional Conversion (% mol.)*
Hydrolysis		
$C_{12}H_{22}O_{11} + H_2O = 2\ C_6H_{12}O_6$	$C_{12}H_{22}O_{11}$	90
Acidogenesis		
$C_6H_{12}O_6 + 2\ H_2O = 2\ C_2H_4O_2 + 2\ CO_2 + 4\ H_2$	$C_6H_{12}O_6$	95
$C_3H_8O_3 = C_3H_6O_2 + H_2O$	$C_3H_8O_3$	99

(continued)

Table 2. *(continued)*

Reactions	Limiting Component	Fractional Conversion (% mol.)*
$C_3H_6NO_2S + 2\ H_2O = C_2H_4O_2 + NH_3 + CO_2 + 0.5\ H_2 + H_2S$	$C_3H_6NO_2S$	100
Acetogenesis		
$C_2H_5OH + H_2O = C_2H_4O_2 + 2\ H_2$	C_2H_5OH	90
$2\ CO_2 + 4\ H_2 = C_2H_4O_2 + 2\ H_2O$	CO_2	100
$C_3H_6O_2 + 2\ H_2O = C_2H_4O_2 + CO_2 + 3\ H_2$	$C_3H_6O_2$	100
$C_4H_8O_2 + 2\ H_2O = 2\ C_2H_4O_2 + 2\ H_2$	$C_4H_8O_2$	100
Methanogenesis		
$2\ C_2H_5OH + CO_2 = 2\ C_2H_4O_2 + CH_4$	C_2H_5OH	100
$C_2H_4O_2 = CH_4 + CO_2$	$C_2H_4O_2$	79
$CO_2 + 4\ H_2 = CH_4 + 2\ H_2O$	H_2	99

* Values obtained from[10]

The hot vinasse enters a heat exchanger (Cooler VS), simulated using the Cooler module where it is cooled with water to 55 °C. The cooled vinasse then flows into the anaerobic digester, simulated using the Conversion Reactor module (Anaerobic Biodigestor). The treated vinasse stream exiting the digester is recirculated back to the digester at a ratio of 0.5. The produced biogas is compressed to 176.5 kPa using the Compressor module (Compressor 1) before entering the desulfurization stage, which is simulated with the Component Splitter module (Desulfurizer). A washing water stream is fed into this stage at a ratio of 6.59 kg washing water per kg of biogas [10], set via the SET-2 module.

For the biomethane production process, the desulfurized biogas is compressed to 1 MPa using a series of compressors (Compressor 2, Compressor 3) and intercoolers (E-100, E-101), maintaining zero pressure drop ($\Delta P = 0$ kPa) across the coolers. The absorption column is simulated using the Absorber module (Absortion Column) with 11 theoretical stages and no pressure drop. The methane-rich stream exiting the absorber passes through a Separator module where the pressure is reduced to 700 kPa to recover residual methane, which is then recirculated to the biogas compression section using the Recycle module (RCY-1). The CO_2-rich water from the separator flows through a throttling valve (VLV-100) to reduce its pressure to 101.3 kPa before entering the desorption column, simulated with another Absorber module (Desortion Column) configured with 7 theoretical stages and no pressure drop. To prevent accumulation of N_2 and O_2, a 1% purge stream is extracted from the water loop [12], with fresh makeup water added as replacement. The regenerated water is pumped back to 1 MPa using the Pump module and recirculated to the absorption column via the Recycle module (RCY-2).

In the final biomethane conditioning stage, residual water is removed using a Component Splitter (Dryer), followed by compression to 20 MPa using two compressor stages (Compressor 4, Compressor 5) with intercooling (E-103, E-104) at zero pressure drop. The washing water and air ratios were maintained at 220.4 kg water/kg CO_2 in biogas and 3.36 kg air/kg CO_2 in biogas respectively, as reported by Cozma et al. [5]. Figure 1 shows the simulation model for biomethane production.

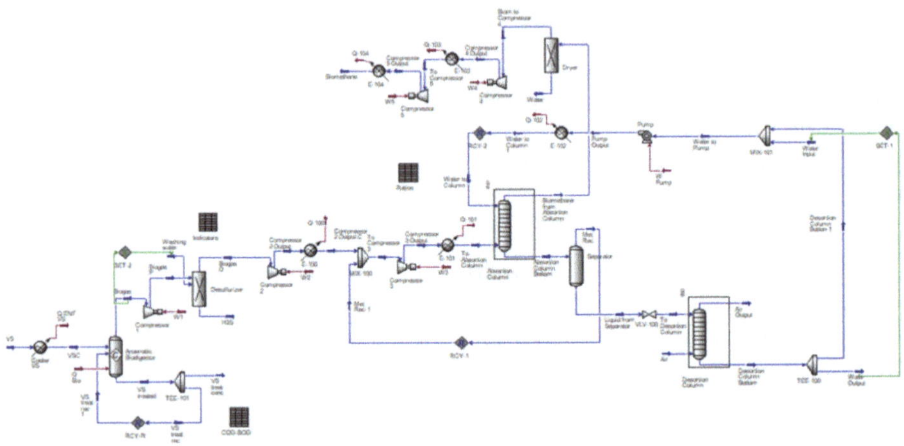

Fig. 1. Simulation Model of Biomethane Production

For the biohydrogen production process simulation, the Cutter module (CUT-100) was used to switch property packages from NRTL-Peng-Robinson to Peng-Robinson only. Cooler modules (E-102, E-103) simulated interstage cooling (to 20 °C) between compression stages, which were modeled using the Compressor module (Compressor 2, Compressor 3) to increase pressure to 2.5 MPa. Water compression (to 2.5 MPa), vaporization, and heating to 909 °C were simulated using the Pump module (Pump) and Heater module (E-101), respectively.

The biogas is heated to 909°C in the Heater module (E-104) before entering the reformer. The reformer (Reformer) and water-gas shift reactors (WSR-1, WSR-2) were simulated using the Equilibrium Reactor module, with the reactions shown in Table 3. Reaction equilibrium data were obtained from Aspen Hysys v10.0's reaction library [13]. Hydrogen separation from residual gas was simulated using the Component Splitter module (PSA). The biohydrogen production simulation model is presented in Fig. 1 (Fig. 2).

Validation of Simulation Models. The validation of the obtained simulation models was carried out by comparing the results from Aspen Hysys v10.0 with data reported in the literature. The absolute or relative error, depending on the property, was calculated in order to determine the difference between the values obtained and the reported values.

Table 3. Chemical reactions incorporated in the biohydrogen simulation model [13]

Chemical Reaction	Module
$CH_4 + H_2O = CO + 3H_2$	Reformer
$CO + H_2O = CO_2 + H_2$	WSR-1, WSR-2

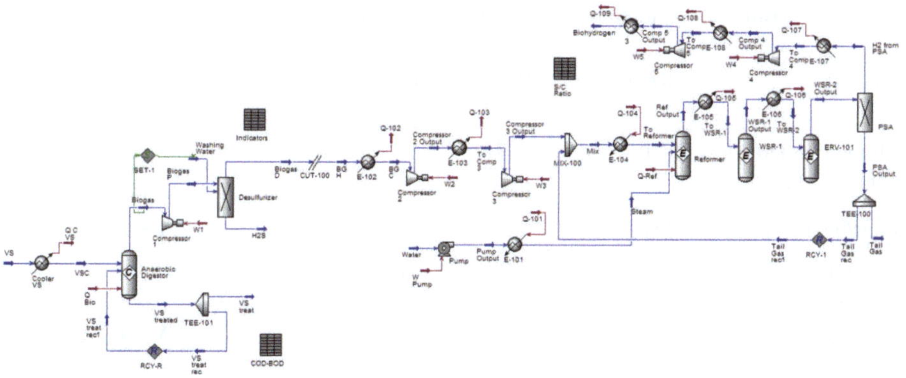

Fig. 2. Simulation Model of Biohydrogen Production

According to Barlas [14], for a model to be considered valid, the error must be less than 5%.

2.3 Simplified Life Cycle Analysis

Emissions associated with the production of biomethane and biohydrogen from vinasse were quantified using a simplified life cycle assessment approach. For both biofuels, direct emissions were considered and calculated by multiplying the leakage flow by the global warming potential factor of 27.9 kg CO_2eq/kg of methane [15]—this was only applied to biomethane, as no leaks occur in biohydrogen production. Combustion emissions from biomethane were considered neutral since it is derived from biomass, while for biohydrogen, they were zero, as its combustion only produces water vapor.

Indirect emissions from energy consumption were determined by accounting for all significant contributions: electrical demand, thermal requirements, and cooling systems. The latter were evaluated using a specific coefficient of performance (COP) of 3.2 for hybrid systems capable of achieving temperatures below 20°C [16]. An emission factor of 0.5 kg CO_2eq/kWh was used for electricity [17], and 0.2 kg CO_2eq/kWh was applied for heating, assuming natural gas usage [15]. For biohydrogen, avoided emissions from residual gas combustion for energy production were also calculated.

The final emission factor for each biofuel was obtained by normalizing total emissions (sum of direct, indirect, and avoided emissions) relative to hourly production. Results were expressed in both kg CO_2eq/kg of fuel and energy basis (kg CO_2eq/MJ), using the lower heating value of each fuel obtained from Aspen Hysys simulation results.

3 Results Analysis

3.1 Simulation Results

The validation results for the simulation models of biomethane and biohydrogen production from distillery vinasse are presented in Tables 4 and 5, respectively.

Table 4. Validation of the biomethane production simulation model

Parameter	Reference	Reported Value	Aspen Hysys	Absolute Error (%)/Criterion
Volumetric methane content (%)	[2]	>97	98.56	Meets criteria
		95–99		Meets criteria
Water/biogas ratio	[5]	0.2	0.1812	1.88
CO_2 removal efficiency (%)	[5]	99.12	98.20	0.92
Methane losses (%)	[2]	<2	0.1918	Meets criteria

The analysis confirms the model's validity, demonstrating full compliance with established criteria while maintaining all errors below 5% threshold [14]. The biomethane compression energy consumption was 0.134 kWh/Nm^3 (similar to 0.11 kWh/Nm3 [18]), while air consumption was 2.29 Nm3 air/Nm^3 biogas (comparable to 2.2 Nm^3 air/Nm^3 [5]). The process yielded 256 kg/h biomethane from 17,800 kg/h vinasse, consuming 120,400 kg/h washing water, 1,835 kg/h air, and requiring 190.97 kW electrical power and 1,435.6 kW cooling capacity.

Table 5. Validation of the biohydrogen production simulation model

Parameter	Reference	Reported Value	Aspen Hysys	Absolute Error (%) / Criterion
Methane conversion in reformer (%)	[6]	93.08	94.12	1.04
Steam conversion in reformer (%)		17.37–28.34	19.14	Meets criteria
Dry basis H_2 from reformer (%)		50.56–69.99	54.8	Meets criteria
CO conversion in WGS-1 (%)	[7]	75	71.59	3.41
H_2 molar composition in biohydrogen (%)		99.99	100	0.01

The validation analysis confirms the model's compliance with all established criteria, including the critical 5% error threshold [14]. The biohydrogen process demonstrated an output of 99.58 kg/h biohydrogen and 1,824 kg/h residual gas (usable as fuel) from 17,800 kg/h vinasse, with total energy requirements of 306.57 kW electrical power, 2,749.91 kW cooling capacity, and 2,787.3 kW heating demand.

3.2 Simplified Life Cycle Assessment Results

Table 6 presents the results of the simplified life cycle assessment for both biofuels, while Table 7 compares their emission factors with natural gas and diesel.

Table 6. Results of the simplified life cycle assessment

Parameter	Biomethane	Biohydrogen	Unit of measurement
Fugitive emissions	13.17	-	kg CO_2eq/h
Electricity consumption emissions	95.49	153.29	
Cooling system emissions	224.32	429.67	
Heating system emissions	-	557.46	
Avoided emissions	-	2,2362.83	
Net emissions balance	332.98	−1,222.4	
Emission factor	1.3	−12.28	kg CO_2eq/kg fuel
Energy-based emisión factor	0.027	−0.102	kg CO_2eq/MJ

Biohydrogen has superior environmental performance compared to biomethane, achieving both negative net emissions balance and negative emission factors. This advantage stems from the process's ability to utilize residual streams for energy production.

Table 7. Comparative analysis of emission factors between renewable fuels (biomethane/biohydrogen) and fossil fuels (natural gas/diesel)*

Fuel	kg CO_2eq/kg fuel	kg CO_2eq/MJ
Biomethane	1.3	0.027
Biohydrogen	−12.28	−0.102
Natural gas	50.3	0.069
Diesel	73.3	0.050

* Emission factors for natural gas and diesel were sourced from [17]

Biohydrogen stands out for its emission avoidance capability, reflected in its negative emission factor. While biomethane still shows positive emissions, these remain significantly lower than natural gas and diesel, positioning it as a cleaner alternative.

These technologies are pivotal for sustainability in sugarcane byproduct industries, transforming highly polluting vinasse into energy sources while aligning with circular economy principles and energy transition trends. They enable more efficient waste valorization strategies and could integrate with other industrial processes like power generation or biofertilizer production.

Simulation models prove essential for assessing technical feasibility and operational conditions prior to large-scale implementation, reducing costs and risks. They serve as tools to analyze process efficiency and scalability while evaluating environmental impacts, including a 70-80% reduction in Chemical Oxygen Demand (COD) for treated vinasse [10] and significant greenhouse gas mitigation. These tools allow for optimization of energy integration and waste-to-resource conversion rates before capital investments.

4 Conclusions

Simulation models were developed in Aspen Hysys v10.0 for biomethane and biohydrogen production from distillery vinasse, validated against literature data with errors below 5%. The biomethane process achieved 98.56% methane purity, while biohydrogen reached 100% composition, making it suitable for fuel applications.

The feasibility study confirmed that both processes are technically and environmentally viable. They have the potential to convert industrial waste into renewable energy, with emission factors of 0.027 kg CO_2/MJ for biomethane and -0.102 kg CO_2/MJ for biohydrogen, outperforming fossil fuels. These results underscore vinasse's value as a biofuel feedstock and its role in advancing circular economy practices in the sugarcane industry.

References

1. Parsaee, M., Kiani Deh Kiani, M., Karimi, K.: A review of biogas production from sugarcane vinasse. Biomass Bioenergy **122**, 117–125 (2019)
2. Francisco, A., Lago, T., Faraji, S., Galera, M., Bello, P.M.: From Biogas to biomethane: an in-depth review of upgrading technologies that enhance sustainability and reduce greenhouse gas emissions. Appl. Sci. **14**(6), 2342 (2024)
3. Tuluhong, A., Chang, Q., Xie, L., Xu, Z., Song, T.: Current status of green hydrogen production technology: a review. Sustainability **16**(20), 9070 (2024)
4. Ghasemi, A., Nikafshan, H., Akrami, M.: Biomass-to-green hydrogen: a review of techno-economic-enviro assessment of various production methods. Hydrogen **5**(3), 474–493 (2024)
5. Cozma, P., Wukovits, W., Mamaliga, I., Friedl, A., Gavrilescu, M.: Modeling and simulation of high pressure water scrubbing technology applied for biogas upgrading. Clean Technol. Environ. Policy **17**, 373–391 (2014)
6. Chouhan, K., Sinha, S., Kumar, S., Kumar, S.: Simulation of steam reforming of biogas in an industrial reformer for hydrogen production. Int. J. Hydrogen Energy **46**(53), 26809–26824 (2021)

7. Son Phan, T., Pham Minh, D., Espitalier, F., Nzihou, A., Grouset, D.: Hydrogen production from biogas: process optimization using ASPEN Plus. Int. J. Hydrogen Energy **47**(100), 42027–42039 (2022)
8. Haydary, J.: Chemical process design and simulation: Aspen Plus and Aspen Hysys applications. John Wiley & Sons, USA (2019)
9. Carlson, E.: Don't gamble with physical properties for simulations. Chem. Eng. Prog. **92**(10), 35–46 (1996)
10. Cruz Llerena, A.: Propuestas de mejoras en la Ronera San José basadas en simulación de procesos. Directors: Osney Pérez Ones, Lourdes Zumalacárregui de Cárdenas. Memoria escrita para optar por el título de Máster en Análisis y Control de Procesos. Universidad Tecnológica de La Habana "José Antonio Echeverría". CUJAE (2020)
11. Cruz Llerena, A., Pérez Ones, O., Zumalacárregui De Cárdenas, L., Pérez De Los Ríos, J.L.: Simulación en Aspen Hysys v10.0 de los procesos de destilación de aguardiente y concentración de vinazas, Icidca sobre los derivados de la caña de azúcar **53**(3), 46–55 (2019)
12. Barbera, E., Menegon, S., Banzato, D., D'Alpaos, C., Bertucco, A.: From biogas to biomethane: a process simulation-based techno-economic comparison of different upgrading technologies in the Italian context. Renew.able Energy **135**, 663–673 (2019)
13. Aspen Technology, Inc. 2017. Aspen Hysys v10.0. https://www.aspentech.com
14. Barlas, Y.: Formal aspects of model validity and validation in system dynamics. Syst. Dyn. Rev. **12**(3), 183–210 (1996)
15. Intergovernmental Panel on Climate Change (IPCC): Climate Change 2021: The Physical Science Basis. Contribution of Working Group I to the Sixth Assessment Report of the IPCC. Cambridge University Press (2021)
16. Wang, S.K.: Handbook of Air Conditioning and Refrigeration. McGraw/Hill, New York (2016)
17. Intergovernmental Panel on Climate Change (IPCC): 2021 Refinement to the 2006 IPCC Guidelines for National Greenhouse Gas Inventories: Volume 2. Energy, IPCC (2021)
18. Budzianowski, W.M., Wylock, C.E., Marciniak, P.A.: Power requirements of biogas upgrading by water scrubbing and biomethane compression: comparative analysis of various plant configurations. Energy Convers. Manage. **141**, 2–19 (2017)

Evaluation of Strategies to Optimize Photovoltaic Penetration in Power Networks

Carlos Darío Castrillón Llanes, Raúl Alejandro Carbonell Ramos, Miguel Ángel Enríquez Labrador(✉), and Ariel Santos Fuentefria

Faculty of Electrical Engineering, Technological University of Havana "José Antonio Echeverría", Havana, Cuba
maelabrados@electrica.cujae.edu.cu

Abstract. In the context of the growing demand for renewable energy sources, this study evaluates the impact of photovoltaic solar energy on secondary distribution networks. The Cuban policy, aimed at increasing the use of this generation source, particularly in residential and commercial applications, may lead to issues in circuits where it is connected if solar penetration is high. To mitigate this effect, different strategies can be implemented to enable greater integration. This work compares three control methods: Fixed Voltage, Perturbation and Observation, and Incremental Conductance. The results indicate that the Fixed Voltage method is the most efficient, achieving a maximum generation of 17.8 kWh/day and a voltage variation of $\pm 1.5\%$, keeping the parameters within acceptable limits. While the Perturbation and Observation method shows lower levels of generation and significant variations in current, the Incremental Conductance method, although efficient, presents greater voltage variation. It is concluded that the Fixed Voltage method is the most recommended for integrating high levels of photovoltaic penetration into secondary distribution electrical networks.

Keywords: Operational strategies · photovoltaic penetration · secondary distribution network

1 Introduction

With the constant development of humanity, from the discovery of electricity to the present day, energy consumption has increased drastically. The burning of fossil fuels and the use of nuclear energy are currently the main means of addressing the energy crisis. However, the environmental damage they cause, such as the release of large amounts of CO_2, SO_2, and NO_2, as well as the generation of radioactive waste, and the scarcity of petroleum, which is fundamental for thermoelectric plants, have led to the search for alternative ways of generating electricity. One of the most prominent alternatives has been the use of renewable energy sources, particularly photovoltaic solar energy. This type of energy has several advantages: it is a sustainable and environmentally friendly option, it reduces fossil fuel consumption, its versatility allows implementation in a wide range of locations, and it has a lifespan of around 25 years. Some factors to consider include its high initial investment cost, although costs have significantly decreased in

A. Santos Fuentefria et al. (Eds.): JIFREEEMA 2025, LNNS 1933, pp. 12–25, 2026.
https://doi.org/10.1007/978-3-032-23028-7_2

recent years, and its dependence on sunlight for production [1]. Worldwide, with the goal of fostering economic growth, improving the quality of life for the population, and, most importantly, protecting the environment, energy development has shifted toward renewable sources. Cuba has adopted measures to reduce fossil fuel consumption and promote the use of photovoltaic solar energy. The country has an energy plan that aims to increase the share of renewable sources in electricity generation to 24% by the year 2030 [2]. As part of this strategy, the installation of solar panels in homes, small businesses, and commercial establishments has been encouraged, which may alter the operational dynamics of the networks to which they are connected. Photovoltaic solar energy has a significant impact on electrical distribution networks. The intermittency of solar radiation can cause fluctuations in energy production, potentially destabilizing the grid [3]. However, it can also improve voltage profiles in the grid, leading to better performance and efficiency. Among the main challenges are the need to store generated energy and manage the intermittency associated with this source. The use of smart grids helps mitigate some of these issues, as they allow real-time monitoring and management of changes in energy production and consumption [4]. For high levels of solar penetration, problems may arise in the distribution circuits where the energy is connected. These problems are associated with voltage variations, active power losses, and other factors related to the quality of electricity delivered to consumers. This study develops an algorithm based on an operational strategy for secondary distribution networks, aiming to identify which strategy allows for higher levels of photovoltaic penetration while maintaining grid parameters within the standards set by regulations [5].

2 Materials and Methods

2.1 Circuit Under Study

For the scenario analysis, a secondary distribution circuit is used (Fig. 1), which consists of 8 nodes and 7 lines, serving a total of 19 consumers. It is observed that in four of them, only phase A is used, three use phase B, and twelve use both phases A and B. The voltage values analyzed throughout the day range from a maximum of 122.05 V to a minimum of 115.19 V. The maximum active power demand from the load is 26.33 kW. In this case, the photovoltaic panels are not yet connected. There are total losses equivalent to 11.35% of the total demand. Regarding voltage levels, the circuit is within the favorable zone (Table 1) [6].

Table 1. Voltage zones

Preferred nominal voltage by the manufacturer	Voltage zones at the equipment terminals			
	Favorable Zone		Tolerable Zone	
	Minimum	Maximum	Minimum	Maximum
120	110	125	107	127
120/240	110/220	125/217	107/214	127/254

(continued)

Table 1. (*continued*)

Preferred nominal voltage by the manufacturer	Voltage zones at the equipment terminals			
	Favorable Zone		Tolerable Zone	
	Minimum	Maximum	Minimum	Maximum
120/208	114/197	125/217	111/193	127/220
240	210	240	200	250
480	420	480	400	500

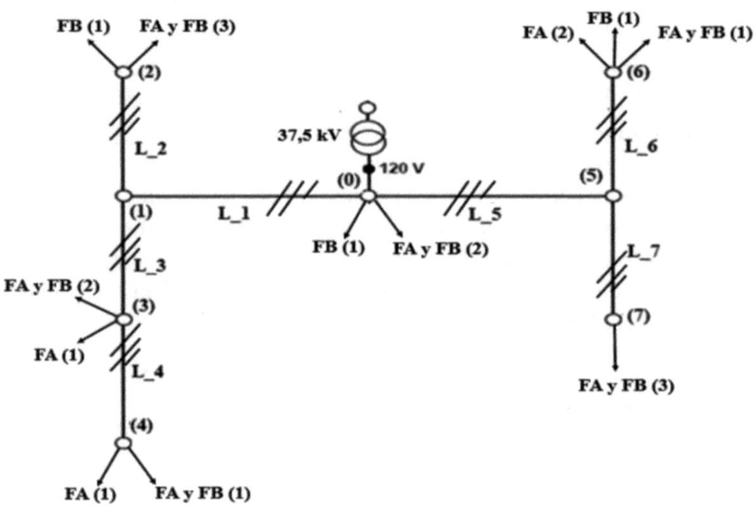

Fig. 1. Three-Phase Radial Secondary Distribution Circuit

2.2 Characteristics of the Photovoltaic Module Used

The selected solar module, being the best-selling in Cuba, consists of 60 solar cells in a 156 mm x 156 mm format, connected in a series-parallel configuration. It is capable of delivering a power output of 250 Wp with a tolerance of $\pm$ 3% under standard radiation conditions [6]. Figures 2 and 3 show the temperature and radiation values in the modules, respectively, for an ambient temperature measured on November 15, 2022, in Havana [6].

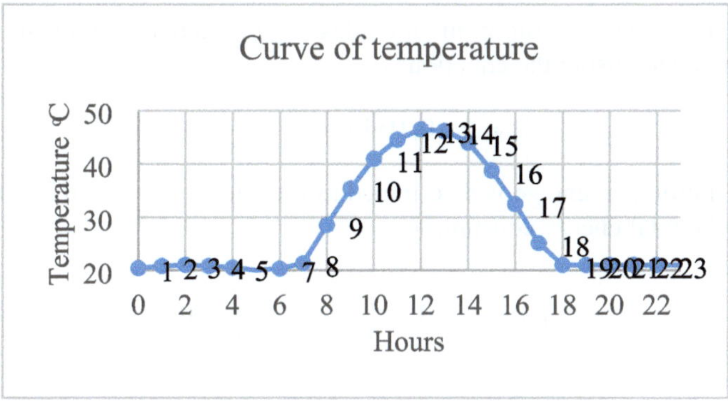

Fig. 2. Temperature on solar module

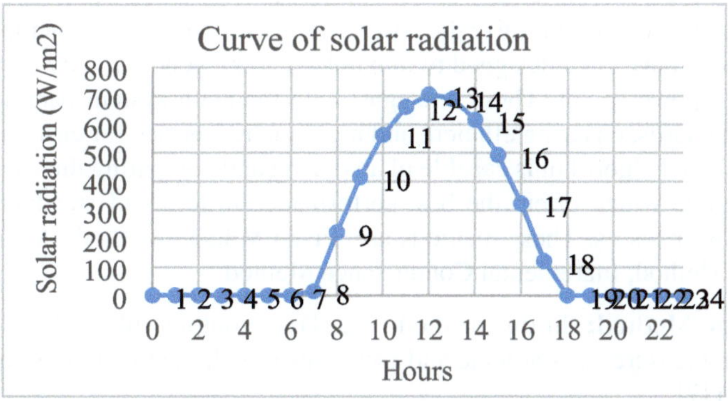

Fig. 3. Radiation on solar module

The power delivered by the PV system is calculated as follows [6]:

$$Psfv_{out} = Psfv \cdot \frac{G}{G_{stc}}(1 + K \cdot (\mathrm{T_{cell}} - \mathrm{T_{stc}})) \cdot N \tag{1}$$

where:

$Psfv_{out}$: Power generated by the photovoltaic system
$Psfv$: Peak power of the photovoltaic module (Wp)
G: Solar radiation
G_{stc}: Standard radiation condition
k: Power correction coefficient due to temperature
T_{cell}: Cell temperature
T_{stc}: Standard module temperature
N: Number of modules

The increase in temperature in the modules (T_{cell}) relative to ambient temperature (T) can be calculated using the equation:

$$\mathrm{T_{cell}} = (0,034 \cdot G - 4) + T \tag{2}$$

The penetration percentage is determined as the energy generated by all the panels compared to the total energy consumption.

$$\%Psfv_{out} = \left(\frac{\sum_{i=1}^{N} Psfv_{out}}{\sum_{i=1}^{19} W} \right) \cdot 100 \tag{3}$$

2.3 Strategies

Consumers have the possibility of installing solar systems in secondary distribution networks; however, it is necessary to take measures to mitigate the potential negative effects this may cause. To optimize the cost of energy production in generation systems, control methods have been designed to maximize energy generation by always seeking the maximum power point. These strategies help maintain system parameters within regulatory standards, ensuring efficient and high-quality service for customers.

The developed algorithm is capable of analyzing and comparing three control methods to determine which offers the best performance in terms of steady-state voltage and active power losses. These methods are: Fixed Voltage Method, Perturbation and Observation Method, Incremental Conductance Method.

Fixed Voltage Method: In this method (Fig. 4), the output voltage $\mathrm{U_{pv}}$ is fixed, and the algorithm measures this voltage and compares it with the maximum power voltage $\mathrm{U_{mpp}}$ (Fig. 4) [7].

$$\Delta U = U_k - U_{k-1} \tag{4}$$

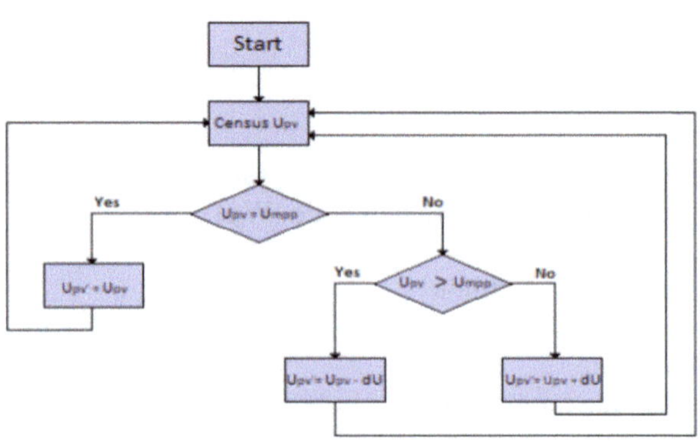

Fig. 4. Python programming algorithm

The voltage must increase in steps of ΔU until the power is as close as possible to the maximum power point. This occurs if the operating point is located in Zone A. If it is in Zone B, the voltage must be decreased in steps of ΔU (Fig. 5). This control method can be implemented by adjusting the current from the photovoltaic module; however, it is recommended to manipulate the voltage, as the objective is to maintain a stable voltage regardless of the type of converter used [8].

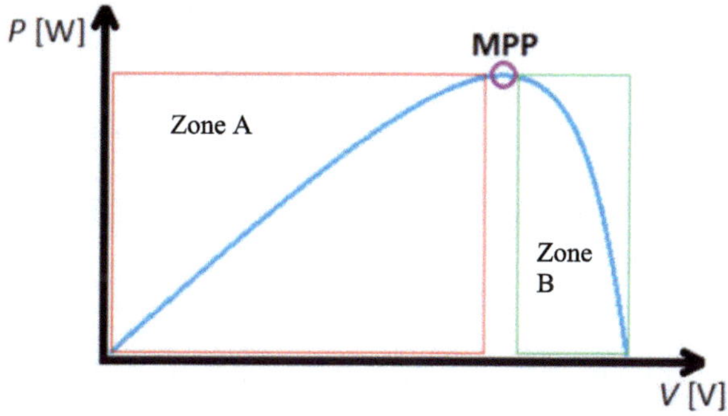

Fig. 5. P-V characteristic of a photovoltaic module.

Perturbation and Observation Method: In the Perturbation and Observation Method (Fig. 6), an analysis of the output power is performed, observing whether there is a variation compared to its previous state. The algorithm processes the comparison and determines whether to increase or decrease the output voltage. The goal is to converge to the maximum power point (MPP) through *n* iterations. The system may exhibit oscillatory behavior (a continuous alternation between the points of lowest and highest power), which would result in efficiency lower than the ideal. Due to abrupt climate changes and minor perturbations in the system, this method may lack precision. The following equation describes the power change that defines the strategy of the method [9]:

$$\Delta P = P_k - P_{k-1} \tag{5}$$

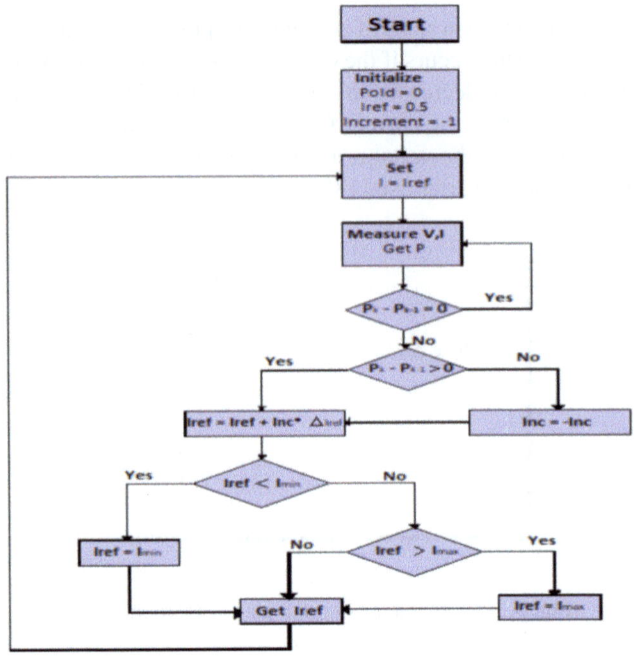

Fig. 6. Python programming algorithm

Incremental Conductance Method: The ease with which equipment or materials allow the passage of current is known as conductance, which is the inverse of resistance:

$$G = \frac{1}{R} = \frac{I}{U} \tag{6}$$

In this method, the extreme value theorem is applied to the P-V curve, where the MPP is found when:

$$\frac{dP}{dU} = 0 \tag{7}$$

Expressing this in terms of voltage and current:

$$\frac{dUI}{dU} = I + U\left(\frac{dI}{dU}\right) = 0 \tag{8}$$

It is observed that the MPP is obtained when the incremental conductance (IC) is the negative of the instantaneous conductance:

$$\frac{dI}{dU} \approx \frac{\Delta I}{\Delta U} \tag{9}$$

The relationship between ΔI/ΔU and I/U is monitored when (Table 2):

Table 2: Relationship between ΔI/ΔU and I/U.

$\frac{\Delta I}{\Delta U} = -\frac{I}{U}$	$U = U_{MPP}$
$\frac{\Delta I}{\Delta U} < -\frac{I}{U}$	$U < U_{MPP}$
$\frac{\Delta I}{\Delta U} > -\frac{I}{U}$	$U > U_{MPP}$

The algorithm (Fig. 7) analyzes voltage and current to calculate both incremental and instantaneous conductance. It then performs a comparison and determines the adjustment in the duty cycle that will be applied to the DC/DC converter, allowing the output voltage to vary and ensuring the delivery of the maximum required power to the load [7].

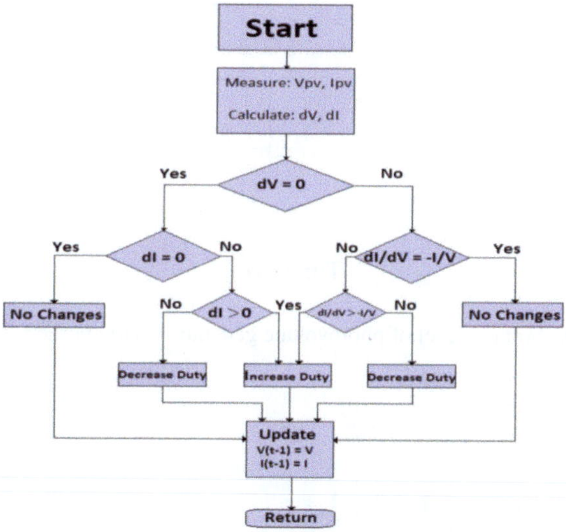

Fig. 7. Python programming algorithm

3 Results

3.1 Fixed Voltage Method

The use of the method in the studied system presents a significant impact of photovoltaic solar penetration on the voltage of the circuit nodes. When these panels generate energy, they must be adjusted to the voltage levels required for integration into the electrical grid (Fig. 8). In this case, a parameter was set to ensure that losses do not exceed 5% of the total system demand, defining a modification limit that the program could implement.

It is highlighted that the voltage level (Fig. 9) showed a variation of ±1.5% compared to the initial conditions, with an increase of 0.9% in the minimum values and 0.14% in the maximum values, reaching 116.265 V and 120.216 V, respectively, maintaining these values within the favorable zone. For a peak hour generation of 2.094 kWh before applying the strategy, an improvement of 51.9% was achieved, equivalent to 1.087 kWh, reaching a value of 3.181 kWh for that hour. Of the maximum generation of 14.63 kWh, an increase of 21.91% was obtained, which corresponds to 3.2 kWh, generating 17.845 kWh per day.

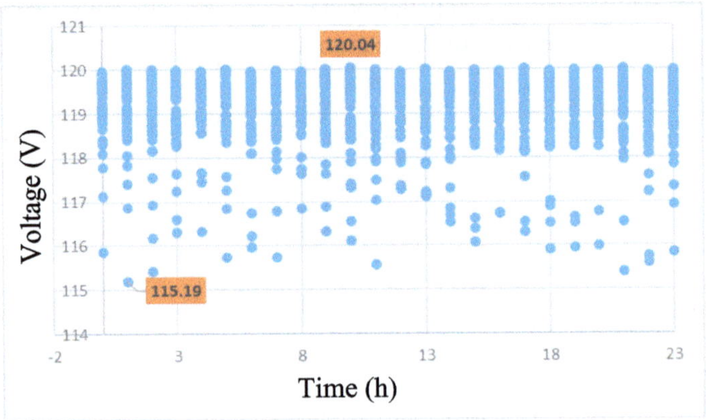

Fig. 8. Voltage level of photovoltaic generation without the method.

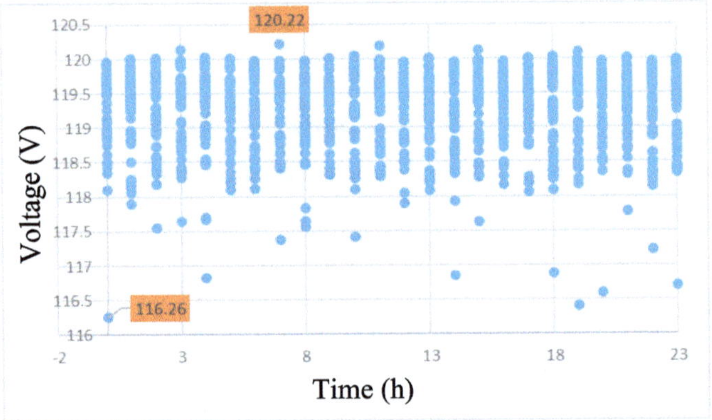

Fig. 9. Voltage level of photovoltaic generation with the method.

3.2 Perturbation and Observation Method

For the strategy analysis (Fig. 10), an optimization limit is set regarding current, with a maximum and minimum value of 30.56 A and 0.4799 A, respectively. Due to these limits, the maximum current decreases by 55%, while the minimum increases by 0.7% (Fig. 11), maintaining voltage levels with negligible variation and achieving losses of less than 5% of the total demand.This approach allows for a daily generation of 16.01 kWh, which is 1.38 kWh higher than the figure without the applied strategies. During peak hours, an increase of 50.45% is observed, raising the initial 2.094 kWh to 3.15 kWh.

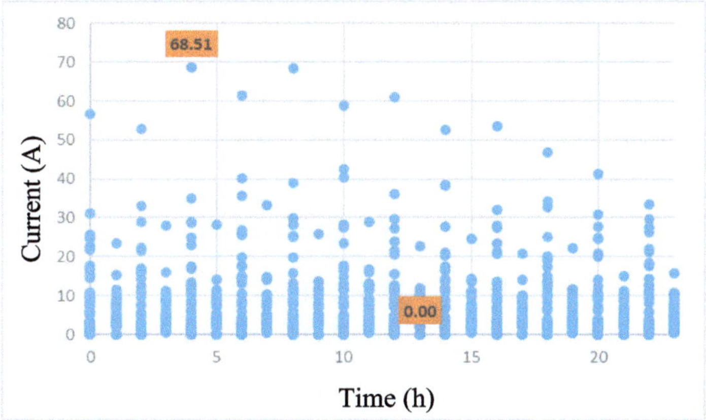

Fig. 10. Current level of photovoltaic generation without the method.

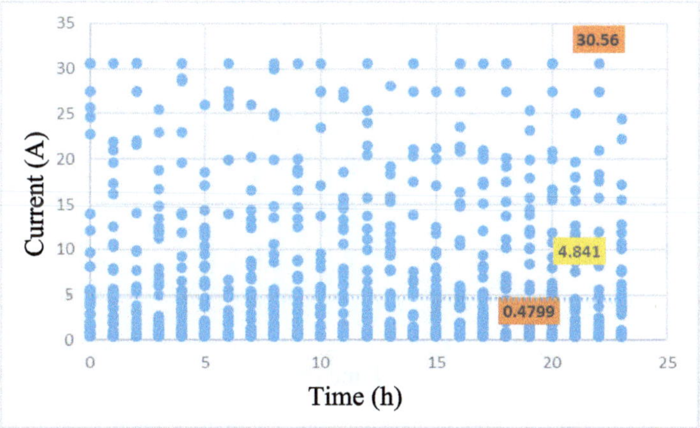

Fig. 11. Current level of photovoltaic generation with the method.

3.3 Incremental Conductance Method

For this method, a 5% variation in the duty cycle is considered to maintain voltage values within the favorable zone and keep losses below 5% of the total demand. With a voltage variation of ±4.5%, the resulting maximum and minimum voltages are 125 V and 112 V, respectively, remaining within the favorable zone.

Regarding the energy generated (Fig. 12), an improvement was observed after applying the method (Fig. 13) compared to its previous state, increasing generation by 49.4% during peak hours, which corresponds to 1.03 kWh. Additionally, the daily generation improved from 14.63 kWh to 17.31 kWh.

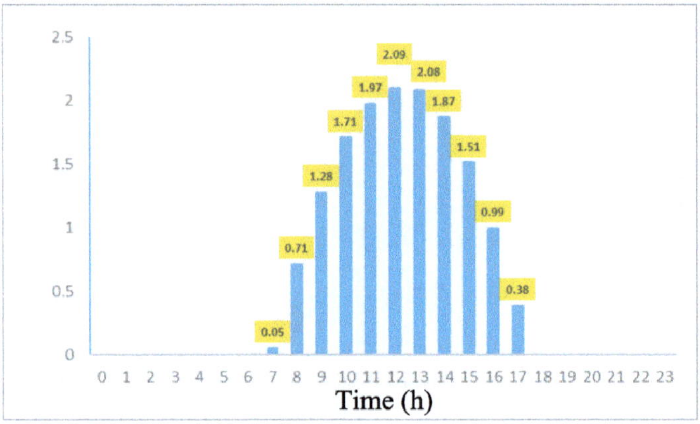

Fig. 12. Photovoltaic generation without the method.

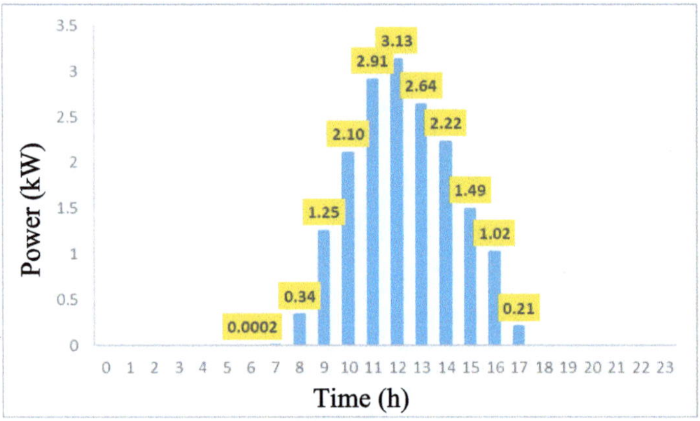

Fig. 13. Photovoltaic generation with the method.

3.4 Analysis and Comparison of Results

The oscillatory behavior in the Perturbation and Observation Method arises due to the way it determines adjustments in the photovoltaic system's operating point. The algorithm perturbs the system by slightly increasing or decreasing the voltage and then observes the resulting change in power. If the power increases, the next perturbation is made in the same direction; if it decreases, the perturbation is reversed. However, due to the iterative nature of this method, the system may oscillate around the maximum power point (MPP), never settling precisely at an optimal point. These oscillations can be exacerbated by sudden changes in environmental conditions, such as solar irradiance or temperature fluctuations, leading to instability. In contrast, the Fixed Voltage Method eliminates this oscillatory behavior because it is programmed to maintain a stable voltage near the MPP rather than continuously adjusting it based on power feedback. By fixing the voltage at an optimal level, the method prevents the small but continuous variations inherent in P&O. Your explanation aligns well with this distinction: while the P&O method primarily focuses on keeping the current within defined limits at 120V, the TF method ensures a more direct approach toward a stable maximum power value, minimizing unnecessary fluctuations. Additionally, because P&O continuously perturbs the system, it may result in inefficiencies due to power losses associated with constant adjustments. Meanwhile, TF achieves a more controlled and predictable integration into the electrical network, making it more suitable for high photovoltaic penetration scenarios.

The difference in over-voltages between the Fixed Voltage and Incremental Conductance methods, despite their similar photovoltaic penetration percentages, is primarily due to the nature of their respective control strategies. In the TF method, the voltage is fixed at a value close to the maximum power point (MPP). This ensures that all operating conditions remain centered around this stable voltage, leading to minimal fluctuations. Since the system does not continuously adjust voltage in response to power feedback, variations are constrained within a narrow range, keeping over-voltages under control. On the other hand, the IC method constantly compares the incremental conductance $\left(\frac{\Delta I}{\Delta U}\right)$ with the negative of the instantaneous conductance $\left(-\frac{I}{U}\right)$. Depending on this comparison, the algorithm determines whether the voltage is above or below the MPP and then adjusts the duty cycle accordingly. This continuous adjustment introduces more pronounced voltage variations because the system is always responding dynamically to changes. As a result, the IC method exhibits greater voltage instability compared to TF, even though both methods lead to similar photovoltaic penetration levels.

Initially, the circuit, after integrating the panels but without applying any method, maintained a maximum voltage of 120.043 V and a minimum of 115.191 V, staying within the favorable zone, with a daily generation of 14.63 kWh and a photovoltaic penetration of 100.85%. After applying the three strategies, an improvement in generation levels is observed (Fig. 14), while the voltage remains within the favorable zone. The Fixed Voltage (TF) method achieved the highest peak-hour and daily generation among the three strategies, with only a 1.5% voltage variation. The photovoltaic penetration for this case was 106.128%, making it the highest of the three. The Perturbation and Observation (P&O) method resulted in the lowest generation, both in peak hours and daily, although its voltage variation was negligible. However, its current fluctuations were significant, which could lead to system imbalances and over-voltages, making it the least recommended approach. The penetration for this strategy was 105.504%, the lowest of the three, though the difference from the TF method was only 0.624%. The Incremental Conductance (CI) method exhibited the most pronounced voltage variation (4.5%), with values ranging from 125V to 112V. Although still within the favorable zone, these values were close to the limit, leaving the system vulnerable to exceeding threshold levels with small variations. Generation levels for peak hours and daily usage were second best, closely approaching those of the TF method, with a photovoltaic penetration of 106.114%. However, due to the voltage stability concerns, it is considered a favorable strategy but not the most recommended. Among the three strategies, the TF method stands out as the best choice, delivering superior performance in generation, penetration, and voltage variation control. It provides the most stable and optimized results in all evaluated aspects (Table 3).

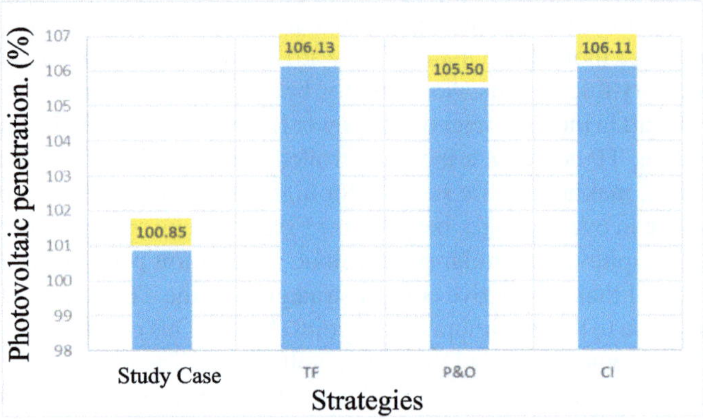

Fig. 14. Photovoltaic penetration level.

Table 3. Comparison between strategies

Strategy	Variation	Penetration level	Conclusions
Fixed Voltage	±1.5% U y + 21.91% P	106.18	Best overall strategy. Highest generation, best penetration, and most stable voltage
Perturbation and Observation	−55% I y + 9.44% P	105.503	Least recommended. Lower generation and significant current fluctuations, which may lead to system imbalance
Incremental Conductance	±4.5% U y + 18.27% P	106.113	Effective but risky. High generation but voltage variations are near critical limits, making it less stable than TF

4 Conclusion

The Fixed Voltage (TF) method proves to be the most effective approach for integrating photovoltaic energy into secondary distribution networks. The study highlights how voltage stability plays a crucial role in ensuring optimal performance of solar power systems, as excessive fluctuations can lead to system inefficiencies, grid instability, and increased losses. By maintaining a ±1.5% voltage variation, the TF method ensures that the system operates within regulatory thresholds while maximizing photovoltaic penetration (106.128%) and achieving a daily energy generation of 17.8 kWh. The comparison with alternative methods further reinforces the advantages of TF: The Perturbation and Observation (P&O) method, despite being widely used in real-world applications, was found to be less efficient, with notable variations in current. These inconsistencies can contribute to electrical imbalances and unwanted voltage surges, making it less suitable for large-scale photovoltaic integration. The Incremental Conductance (IC) method, while capable of achieving high energy generation (106.114% penetration), exhibited the highest voltage variation (±4.5%). This factor introduces potential risks, as the system is more susceptible to instability when facing external perturbations such as weather changes or load fluctuations. The contributions lie in its quantitative assessment of photovoltaic integration strategies, providing empirical evidence that supports the TF method as the best solution. By maintaining low voltage fluctuations, the strategy ensures grid compliance, long-term stability, and enhanced energy efficiency—key factors for sustainable solar energy deployment. These findings not only help optimize photovoltaic generation but also establish critical parameters for future smart grid development and renewable energy management strategies.

References

1. Alfonso, J.: Ventajas y desventajas de la energía fotovoltaica. SunFields Europe. https://www.sfe-solar.com/noticias/articulos/ventajas-y-desventajas-energia-fotovoltaica/. Last accessed 04 Sep 2024
2. Castro, R.: Representaciones diplomáticas de Cuba en el exterior. Cuba Minrex. https://misiones.cubaminrex.cu/es/articulo/discurso-pronunciado-por-el-general-de-ejercito-raul-castro-ruzoctavo-periodo-ordinario-de. last accessed 17 Dec 2016
3. Quiñonez, E.: Energía fotovoltaica en redes eléctricas. Inspenet. https://inspenet.com/articulo/energia-fotovoltaica-en-redes-electricas/?form=MG0AV3. Last accessed 18 Oct 2023
4. Endesa: Transformación y digitalización de redes de distribución. Endesa. https://www.endesa.com/es/proyectos/todos-los-proyectos/transicion-energetica/transformacionydigitalizacionredesdistribucion#:~:text=Esta%20transici%C3%B3n%20energ%C3%A9tica%20va%20acompa%C3%B1ada%2de%20dos%20grandes,el%C3%A9ctrico%2C%20y%20nuevos%20. (endesa.com in Bing). Last accessed 10 June 2021
5. Villavicencio, J.: Comportamiento de las redes de baja tensión ante la presencia de generación fotovoltaica. Tecnia p. 7 (2021)
6. Osorio, O.: Algoritmo de optimización en Python para analizar el impacto de generación fotovoltaica y vehículos eléctricos en redes de distribución secundaria. p. 12 (2023)
7. Aguirre, O.: Determinación del punto de máxima potencia de sistemas fotovoltaicos mediante el uso del algoritmo de conductancia incremental. Universidad Politécnica Salesiana, Quito (2023)
8. Duarte, J.: Modelado y simulación de un algoritmo de control de potencia de sistemas de generación fotovoltaicos basado en la técnica P&O - MPPT. Universidad Industrial de Santander, Santander (2015)
9. Pongsakor, T.: Maximum power point tracking using fuzzy logic control for photovoltaic systems. In: IMECS 2011, Hong Kong, pp. 1–5 (2011)

University Management Model Based on Vulnerability and Risk Approaches Oriented Toward Sustainability

Carlos Julio Vinces Solórzano[1](✉), Sebastiana del Monserrate Ruiz Cedeño[1], Jimmy Manuel Zambrano Acosta[1], Daniel Alfonso Robaina[2], and Ariel Racet Valdés[2]

[1] Universidad Técnica de Manabí, Portoviejo, Ecuador
superpeto-1@hotmail.com

[2] Universidad Tecnológica de La Habana "José Antonio Echeverría", La Habana, Cuba

Abstract. This article presents a comprehensive university management model based on a vulnerability-and-risk approach, designed to promote institutional sustainability. The growing challenges posed by climate change, socioeconomic instability, and energy crises provide sufficient grounds to propose a management framework that integrates key components, including the environment, a vulnerability perspective, a management approach, resilience strategies, institutional governance, and a monitoring system. The model adopts a systemic, preventive approach, enabling the identification of exposure, sensitivity, and adaptive capacity factors within the university context. The methodology is based on a descriptive and proactive analysis of interrelated variables, drawn from empirical evidence and an updated literature review. It is concluded that this model serves as a tool to strengthen institutional resilience and promote an organizational culture oriented towards sustainable development.

Keywords: Sustainable university management · institutional resilience · vulnerability · risk · sustainability · higher education · management model · climate change · governance

1 Introduction

Contemporary universities face a complex interplay of environmental, economic, and social pressures that threaten their sustainability. As key agents in sustainable development, these institutions must strengthen their adaptive and responsive capacities to address emerging risks. Traditional management approaches focused solely on efficiency or quality are no longer sufficient to ensure long-term viability. In this context, it becomes essential to integrate frameworks that address institutional vulnerability, risk scenarios, and resilience capacities, particularly in regions prone to climate threats or energy crises [1].

A. Santos Fuentefria et al. (Eds.): JIFREEEMA 2025, LNNS 1933, pp. 26–32, 2026.
https://doi.org/10.1007/978-3-032-23028-7_3

This article proposes a sustainability-oriented university management model that articulates nine key components: environment, vulnerability perspective, management approach, resilience strategies, governance, environmental regulatory framework, resilient infrastructure, clean technologies, and monitoring systems. The model acknowledges the need to shift toward more adaptive, preventive, and sustainable institutional structures capable of anticipating and mitigating risks without compromising their educational and social mission [2–5].

Sustainability in higher education has evolved from initiatives focused on environmental management to comprehensive approaches to organizational resilience. Several studies emphasize the urgency of incorporating vulnerability and risk analysis into university strategic planning [6].

The concept of institutional vulnerability refers to an organization's exposure to harm during crises, depending on its sensitivity and response capacity [7]. In the university context, this involves analyzing energy dependence, infrastructure quality, human resource training, and the robustness of governance structures [8].

On the other hand, institutional resilience is associated with a university's ability to recover, adapt, and transform in the face of disruptions [1, 2]. Models such as adaptive management [2–5] and risk-based management [9] have proven effective in complex and dynamic contexts.

2 Method

This study employs a descriptive–propositional methodological approach, grounded in documentary analysis and conceptual modeling. A triangulation strategy was employed, combining a review of scientific literature (2020–2025), an analysis of international best practices, and accumulated experience in managing higher education institutions in Ecuador and Cuba.

3 Results

The University Management Model based on Vulnerability and Risk Approaches for Sustainability (MGU-EVRS, by its initials in Spanish) is structured around nine interrelated components:

1. Environment: External conditions influencing the university's operations (climate change, regulations, political and economic context) [10, 11].
2. Vulnerability perspective: Analysis of the institution's exposure, sensitivity, and adaptive capacity in the face of threats [1], considering financial conditions, information and communication technologies (ICTs) development, and academic performance.
3. Management approach: Sustainability-oriented, integrating strategic planning, process management, and a preventive approach [12], characterized by proactive decision-making, strategic integration, and continuous improvement.
4. Resilience strategies: Institutional measures to prevent, respond to, and recover from crises (contingency planning, training, innovation) [1], including innovation in university management, student enrollment levels, and especially social inclusion.

Figure 1 shows a graphical representation of the proposed model and its nine components.

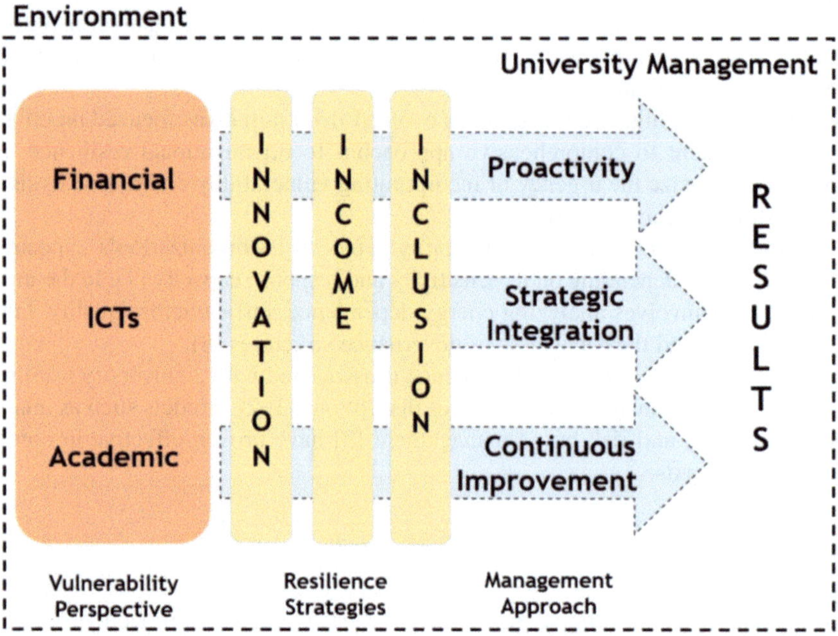

Fig. 1. University management model based on vulnerability and risk approach.

The main goal of the MGU-EVRS is to strengthen universities' institutional resilience by systematically integrating environmental, social, economic, and governance variables into planning, decision-making, and evaluation processes. This approach aims to proactively anticipate, mitigate, and adapt to emerging risks while fostering a sustainable, equitable, and transformative higher education system.

4 Principles of the Model

- Comprehensive sustainability: The model promotes university development that balances economic, social, and environmental factors in alignment with the Sustainable Development Goals (SDGs) [8].
- Systemic approach: It views the university as a complex system in which internal and external components dynamically interact, enabling a holistic understanding of its management [2–5].
- Prevention and adaptation: The model prioritizes anticipating risks and implementing adaptive measures in uncertain scenarios, rather than reactive responses [1].
- Institutional resilience: It fosters universities' capacity to withstand, recover from, and transform in the face of disruptions, ensuring the continuity of their educational mission [2–5].

- Participation and democratic governance: The model encourages active participation by all university stakeholders in management processes, underpinned by a commitment to sustainability [8].
- Technological and ecological innovation: It stimulates the adoption of clean technologies and innovative solutions that help mitigate environmental impact and improve institutional efficiency [13–15].
- Transparency and continuous improvement: The model establishes monitoring, evaluation, and feedback mechanisms to ensure its effectiveness and ongoing evolution [16].

5 Fundamental Premises of the Model

1. The university is a strategic actor in sustainability and must lead by example in the responsible management of its resources and environment.
2. Institutional risks are not merely random events but the result of accumulated vulnerabilities that can and must be managed.
3. Managing vulnerability is a necessary condition for sustainability, as it reduces exposure and sensitivity to internal and external threats.
4. The university environment is dynamic and uncertain, requiring flexible, adaptive, and proactive organizational capabilities to manage change.
5. Institutional culture is key to the model's success, as it demands a transformation in the values, attitudes, and practices of all stakeholders.
6. There is no sustainability without equity; therefore, the model incorporates the social dimension of resilience and intergenerational justice as transversal principles.

The proposed model goes beyond traditional approaches focused exclusively on educational quality or operational efficiency. The model introduces sustainability as a central axis, examined through the lenses of risk and resilience. The relationships among the nine components allow for a comprehensive assessment of institutional performance under changing conditions. Furthermore, its systemic approach encourages interdepartmental participation and transparency in decision-making. However, its implementation requires political and institutional commitment, as well as specialized technical capabilities.

An ideal behavioral pattern is developed to assess university performance, as shown in Table 1.

Table 1. Description of the behavior of the variables in the MGU-EVRS model

Variable	Level 0 Critical (0%)	Level 1 Low (25%)	Level 2 Intermediate (50%)	Level 3 Advanced (75%)	Level 4 Ideal (100%)
Environment	The university ignores changes in the environment and does not assess external risks	External threats are recognized but not integrated into planning	Basic environmental analysis exists, but with no integrated actions	Environmental analysis guides strategic decision making	Predictive analysis and agile response to the environment; dynamic planning
Vulnerability	Vulnerabilities are neither identified nor assessed	Partial and reactive diagnosis during crises	Periodic diagnosis, but no sustained corrective actions	Action is taken on critical points; follow-up mechanisms are in place	Integrated system for continuous vulnerability assessment and reduction

(*continued*)

Table 1. (*continued*)

Variable	Level 0 Critical (0%)	Level 1 Low (25%)	Level 2 Intermediate (50%)	Level 3 Advanced (75%)	Level 4 Ideal (100%)
Management Approach	Fragmented management with no strategic planning	Incipient planning with no proactive approach	Partial process-based management, with limited linkage and integration with sustainability	Integrated planning with a focus on integration and sustainability	Strategic management with a systemic approach, continuous improvement, and resilient sustainability
Resilience Strategies	No contingency plans or response capacities exist	Basic plans exist but lack resources or updates	Response strategies are defined but poorly articulated	Operational and training strategies in place, with defined leadership	Institutional response capacity, with simulations, continuous adaptation, and innovation

6 Discussion

The proposed university management model is based on vulnerability and risk approaches geared toward sustainability, representing a necessary evolution in higher education management frameworks that integrates dimensions that have traditionally been addressed in a fragmented or reactive manner. Unlike traditional models, this proposal incorporates a holistic, anticipatory, and systemic perspective, where the analysis of vulnerability, risk, and resilience is positioned as the central axis of institutional management.

The articulation of these four variables enables a comprehensive understanding of universities as open and complex systems, exposed to changing dynamics in economic, political, environmental, and social contexts. This approach underscores the need to adopt organizational governance models that integrate resilience as a dynamic capacity to address multiple, simultaneous threats in uncertain environments.

From this perspective, the MGU-EVRS transcends the traditional view that considers risk solely as a threat, promoting a strategic vision in which universities assume an active role in identifying, mitigating, and adapting to adverse scenarios. Furthermore, the model integrates principles of comprehensive sustainability and intergenerational justice, consistent with the Sustainable Development Goals (SDGs), emphasizing that university sustainability cannot be dissociated from social equity and democratic participation.

Moreover, the variable of institutional resilience strategies highlights the importance of incorporating technological and organizational innovation not only as a response to crises but as a structural part of a resilient organizational culture, open to continuous learning and transformation.

However, the effective implementation of the model requires overcoming institutional limitations related to organizational culture, resistance to change, and the availability of specialized technical capabilities. Furthermore, robust monitoring and evaluation systems are necessary to enable universities to continuously measure their levels of vulnerability, resilience, and sustainability, thereby ensuring continuous improvement.

In this context, the MGU-EVRS constitutes a strategic tool not only for university management but also to guide public higher education policies that are more coherent with current global challenges, contributing to the consolidation of resilient, sustainable, and socially responsible universities.

7 Conclusions

The MGU-EVRS model constitutes a highly valuable tool for transitioning toward a sustainable, resilient, and adaptive university. Its application enables:

- The early identification of risk and vulnerability factors.
- The formulation of adaptation and mitigation strategies.
- The improvement of institutional governance and resource use efficiency.
- Compliance with the SDGs and the demands of the contemporary environmental and social context.

It is recommended that the model be empirically validated across universities of diverse typologies, and that self-assessment and institutional audit tools based on the model be developed.

References

1. Farfán, J.F., Delgado, R., Farfán, D.E.: Educación ambiental, currículo, estrategias y políticas para la sostenibilidad: una revisión sistemática. Revista Alfa **8**(23), 576–592 (2024). https://doi.org/10.33996/revistaalfa.v8i23.287
2. Van Zanten, J.A., Van Tulder, R.: Improving companies' impacts on sustainable development: a nexus approach to the SDGS. Bus. Strateg. Environ. **30**(8), 3703–3720 (2021)
3. Ungar, M., Theron, L., Höltge, J.: Multisystemic approaches to researching young people's resilience: discovering culturally and contextually sensitive accounts of thriving under adversity. Dev. Psychopathol. **35**(5), 2199–2213 (2023)
4. Zelčāne, E., Pipere, A.: Finding a path in a methodological jungle: a qualitative research of resilience. Int. J. Qual. Stud. Health Well-Being **18**(1), 2164948 (2023)
5. McCormick, K.C., Mendle, J.: Growing pains: the history of human development and the future of the field. Child Dev. Perspect. **19**(3), 179–185 (2025). https://doi.org/10.1111/cdep.12541
6. Grimalt-Álvaro, C., Valls, C., Cebrián, G., Marqués-Molías, L.: An exploration of students' attitudes towards STEM and climate change: a cluster analysis study with a gender perspective. EJSTEME **10**(1), 02 (2025). https://doi.org/10.20897/ejsteme/16206
7. Klasa, K., Trump, B.D., Dulin, S., Smith, M., Jarman, H., Linkov, I.: A resilience augmented approach to compound threats and risk governance: a systems perspective on navigating complex crises. Environments **12**(2), 64 (2025). https://doi.org/10.3390/environments12020064
8. UNESCO. Educación para el desarrollo sostenible: hoja de ruta. París, UNESCO (2020). https://unesdoc.unesco.org/ark:/48223/pf0000374896
9. ISO. ISO 31000:2018 Risk Management Guidelines. Geneva: International Organization for Standardization (2018). https://www.iso.org/standard/65694.html
10. Alireza, Z., Razzaqul, A., Rashid, A.A., Mohammed, A.A.: Cleaner Eng. Technol. **18** (2024). https://doi.org/10.1016/j.clet.2024.100722
11. Abdelsalam, E., Almomani, F., Alnawafah, H.: Triple-renewable energy system for electricity production and water desalination. Environ. Sci. Pollut. Res. **30**, 98895–98906 (2023)
12. ISO. ISO 14001:2015/Amd 1:2024 Environmental management systems. Geneva, International Organization for Standardization (2024). https://www.iso.org/standard/88209.html

13. Espin Ruiz, G.P., Salgado, R.I.M., Timbila Crespo N.M., Riera Salazar, A.C., Gaibor Barragan, C.A.: Educación para la Sostenibilidad: proyectos Transversales sobre medio Ambiente y Responsabilidad Social. Ciencia Latina **9**(2), 905–919 (2025). https://doi.org/10.37811/cl_rcm.v9i2.16896
14. Bedoya, N.: Investigación Educativa para el Desarrollo Sostenible. MACSD Conf. Proc. 8(3), 9170–9185 (2024). https://doi.org/10.37811/cl_rcm.v8i3.12068
15. Gutierrez Cuesta, R., González Jiménez, C.: Impacto de las tecnologías de la información y comunicación en el aprendizaje de las matemáticas en la educación superior. Ciencia Latina **9**(2), 1094–1114 (2025). https://doi.org/10.37811/cl_rcm.v9i2.16934
16. Mhlolo, M.: A broken promise for gifted learners: re-examining the national strategy for mathematical sciences in South Africa two decades later. Creat Ed **13**, 803–820 (2022). https://doi.org/10.4236/ce.2022.133053

Use of Zeolites in the Reduction of Impurities in Vegetable Frying Oils for Biofuels Synthesis

Marilín García Díaz[1(✉)], Taimí Noa Almeida[1], Leandy Mendoza Navarro[1], Yanet Sariego Toledo[1], José Gandón Hernández[1], Carlos M. Cordoví[1], and José R. Correa[2]

[1] Department of Chemical Engineering, Faculty of Chemical Engineering, Technological University of Havana "Jose Antonio Echevarría", CUJAE. 114 St. 11901, Marianao, 19390 Havana, Cuba

mgarcia@quimica.cujae.edu.cu

[2] Department of General and Inorganic Chemistry, Faculty of Chemistry, University of Havana, Zapata St., 10400 Havana, Cuba

Abstract. Used vegetable oils in frying processes constitute a waste that, if not managed correctly, results in environmental damage and deterioration of human health. When used as raw material in the preparation of biofuels, the oil has to purify. The aim of this research was to study the purification process of used cooking oil using natural zeolite as an absorbent material. The physicochemical properties of the oil used were determined. Zeolite was analyzed using X-ray diffraction, Raman Spectroscopy and Scanning Electron Microscopy. A multilevel factorial experimental design was planned for the first adsorption stage. The operational variables investigated were temperature (80, 100, 120 °C), particle diameter (0.315, 0.630, 1 mm), and zeolite ratio (m/m) (5, 10%). The osrder of the reaction was determined from the kinetic study. It was concluded that, to obtain minimum absorbance values, which imply greater impurity removal, it is necessary to work at low temperatures, with smaller particle diameters and a higher proportion of zeolite. For a second absorption stage, varying the proportion of zeolite showed a decrease in absorbance values. Homogeneous transesterification reactions, with both, unpurified and purified oil in two stages had performed and a higher yield with the purified oil was obtained.

Keywords: used soybean oil · natural zeolite · physical adsorption

1 Introduction

Nowadays a large amount of the world production of vegetable oils is intended for use in frying, with a tendency to increase due to the rise of fast foods (Yagüe, 2003; Tacias-Pascacio et al., 2016; Thushari and Babel, 2022). Frying is a process of immersing food in hot oil, at high temperatures (175−185 ºC), where the oil acts as a heat source, producing rapid and uniform heating of the food, providing texture and flavor. In this sense, soybean, sunflower, palm, corn, canola and olive oils are the most used in this

A. Santos Fuentefria et al. (Eds.): JIFREEEMA 2025, LNNS 1933, pp. 33–47, 2026.
https://doi.org/10.1007/978-3-032-23028-7_4

culinary practice (Suaterma, 2011; Ahmad et al., 2021). During frying, the oil is subjected to chemical and physical modifications as a result of its interaction with water and other components of the food, which brings about deterioration as the time of use increases. As a consequence of this interaction, a large number of complex reactions take place, such as hydrolysis, oxidation and polymerization, which result in a decrease in the nutritional quality of the food and an increase in the formation of toxic compounds in the oil. Such toxic substances migrate to the food and therefore, harmful effects for health (Yagüe, 2003; Bansal et al., 2010; Rivera et al., 2014; Cabezas-Zábala et al., 2016; Ahmad et al., 2021; Erickson et al., 2022; Licata, 2023).

On the other hand, used cooking oil (UCO) involves problems for the environment. Due to hydrolysis, oxidation and polymerization reactions, toxic substances are formed that accumulate in the environment and negatively influence the health of human beings and reduce the biota (Rahayu et al., 2021). The chemical transformation of used soybean oil into esters of fatty acids and glycerin, through heterogeneous transesterification, is a solution that would help mitigate the problems that they generate both in health and in their discharge into the environment (Thushari and Babel, 2022). Heterogeneous catalysts and especially solid state catalysts are promising when replacing homogeneous (conventional) catalysts in the transesterification reaction, since they are less corrosive, easier to handle, reusable and generate less amount of toxic waste (Cabello et al., 2017; Mansir et al., 2017; Tobío-Pérez et al., 2021).

In order to accelerate the chemical process adequately, catalysts must display several properties such as activity, selectivity and stability, some of which are the focus of several research (Cabello et al., 2017; Kesić et al., 2016; Algieri and Drioli, 2021; Abdelhameed et al., 2022; Loiola et al. 2022). Catalytic activity is the property of increasing the reaction rate comparing with the same system without a catalyst under identical conditions (Mosquera Canchingre et al., 2022). Generally, catalysts do not remain continuously active for indefinite periods, on the contrary, their activity decreases over time, a phenomenon known as deactivation. Deactivation can be uniform (at all active sites) or selective, in which, the most active sites, those that provide most of the catalyst activity, are preferentially attacked and deactivated.

There are several causes for heterogeneous catalyst decay or deactivate. Fogler (2008) mentions three reasons by which deactivation may occur, they are: poisoning, aging, and contamination or coking. Carballo (2002) classifies the origins of deactivation into three groups: physical causes, impurities in the reactor feed, and reaction products that are deposited on the catalyst surface. On the other hand, Argyle and Bartholomew (2015) state that the deactivation of heterogeneous catalysts, depending on its mechanism, can be caused by: poisoning, fouling, thermal degradation, vapor formation, vapor-solid and solid-solid reactions and wear/crushing. As stated above, poisoning is one of the causes that most influences the deactivation of catalysts. Because of deactivation of catalysts can be present, it is of vital importance that the used oil from the frying processes is subjected to a pretreatment process before the transesterification reaction takes place, in order to help avoid poisoning of the solid catalyst.

Depending on the type and the application selected, several kinds of treatments for oils are used. For example, to refine a vegetable oil there are two methods: physical or chemical or a combination of both, which generally requires the following steps:

degumming, neutralization, bleaching and deodorization. When oil is conditioned for the use in specific processes, it is suggested that only the solid residues and the aqueous phase that may be present should be separated. Rahayu et al. (2021) used sugarcane bagasse activated charcoal for the purification of used cooking oil for the conversion into soap. In the case of recycling the UCO in transesterification reaction, using heterogeneous catalysis, it is not enough the traditional treatment of cleaning. As a consequence of the frying process, compounds that are formed can occupy active sites of the catalyst, and therefore, an adequate proposal for the adsorption process should be considered, which does not appear frequently in the literature.

Zeolites are three-dimensional aluminosilicate materials that possess framework structures with channels and cavities. The main chemical elements present in the zeolite structure are Si, Al and O disposed in tetrahedral network and chains. The radicals are of the general formula $[Si_{n\text{-}p}Al_pO_2]^{p-}$, where n can takes values from 5 to 24 an p from 1 to 5. The negative charge is balanced by counter-ions such as alkaline Na^+, K^+, or alkaline earth Ca^{2+}, Sr^{2+}, Ba^{2+}, among other. The cavities can also contain water molecules and determine the main zeolite properties, thus these materials show molecular sieving capacity, high adsorption property and ion exchange capacity. Zeolites are used for removal of heavy metals, separation of hydrocarbons, gas adsorption and water softening (Milovski and Kononov, 1988; Algieri and Drioli, 2021). In particular the natural zeolite clinoptilolite was reported as used in soil amendment, feeds and water purification, and for crude oil spill removal from sea water (Kalbuadi et al., 2019).

The adsorbent property of zeolite can be avail for the purification of used edible oil from food industry. Thus, the objective of this work is to study the purification process of used cooking oil using natural zeolite as an absorbent material, to minimize catalyst poisoning in the transesterification reaction.

2 Material and Methods

2.1 Materials

The used soybean oil (USO) was supplied by the Pre-form Plant of the Food Production Company (Prodal), Havana, Cuba. Two oil samples with different levels of use were usedand the natural zeolite (ZP) from the Tasajera deposit, supplied by the Center for Research and Project for the Metallurgical Mining Industry (CIPIMM), Havana, Cuba.

2.2 Physicochemical Characterization of USO

The determination of polar compounds was carried out using the Testo 270 device. The equipment was calibrated using reference oil whose total polar compound limit was 6.9% with a 0.5 tolerance level. To perform the measurement, the sample was heated to a temperature of 168 ºC. International standards suggest that the maximum value of the percentage of polar compounds should not be greater than 25% (Suaterna-Hurtado, 2011; Boatella-Riera and Codony, 2000). The density was determined by traditional methods, at room temperature, for which a 25 ml pycnometer was used. The density was determined in triplicate and the arithmetic mean of the three measurements as well as standard deviation were taken as the final result. In order to measure viscosity, a VISCO STAR–R model viscometer was used, which is appropriate for determining viscosities in a range of 20–13,000,000 cP by the rotatory method.

The amount of Free Fatty Acids (FFA) was found by the quantification of Acidity Index (AI) and Degree of Acidity of the oil from the Cuban Norm NC 660:2001 and NC 660, 2021 standard. A solution of 0.1 N of KOH in ethanol was used as a solvent. The used formulas are represented in Eqs. (1) and (2).

$$\mathrm{FFA}(\%_{\,\mathrm{OleicAcid}}) = Vc_{\mathrm{KOH}}M_{\mathrm{Oleic\ Acid}} \Big/ 10\ \mathrm{m_S} \tag{1}$$

$$\mathrm{AI}\left(\mathrm{mg_{KOH}} \big/ \mathrm{g_{Oil}}\right) = VcM_{\mathrm{KOH}} \big/ m_{\mathrm{S}} \tag{2}$$

where V: volume of the titrated potassium hydroxide solution (ml). c_{KOH}: the equivalent concentration of KOH solution (mol/l). $M_{\mathrm{Oleic\ Acid}}$: equivalent molar mass of the oleic acid (282 g/mol) for which the result is expressed. M_{KOH}: molar mass of potassium hydroxide (56.11 g/mol). m_{S}: mass of the sample used (g). Both properties were determined in triplicate and the arithmetic mean of three measurements was taken as the final result. The standard deviation was reported.

2.3 Preparation of Used Soybean Oil

First, the oil was filtered in an MLW model DSE2 vacuum filter, through Whatman 202 filter paper. This procedure was carried out with the aim of eliminating the suspended solids that the oil brings as a result of the frying process.

2.4 Preparation and Characterization of Natural Zeolite

The selected natural zeolite (ZP) exhibits a particle size greater than that required in this research. In order to reduce the particle size, it was ground with a ball mill and sieved. Moisture was removed by heating at 120 °C for 2 h directly on a heating plate. The grinding was carried out with the objective of reducing the size of the particles through the combination of impact and abrasion. The operation is carried out in a planetary ball mill and 150 porcelain balls, with an average diameter of 24.83 mm, were used. The grinding times were adjusted until the desired particle size was reached, which was of 15 min. To obtain the natural zeolite with the necessary particle size, a system of 4 sieves was used, with opening diameters of 2.0; 1.0; 0.63 and 0.315 mm.

For phase and morphological characterization X-ray diffraction and Scanning Electron Microscopy analysis were conducted. The diffractogram of the zeolite sample was recorded using a PW 3050/60 goniometer X-ray diffraction (XRD) instrument from Malvern Panalytical with Cu Kα radiation and a nickel filter, operated at 45 kV and 40 mA. A Tescan Scanning Electron Microscope was used for the SEM under microanalysis conditions, operated at 20 kV, with a 5 mm working distance and a secondary electron detector.

2.5 Pretreatment of Used Soybean Oil

The pretreatment was carried out in a hermetically sealed 100 ml beaker on a heating plate and magnetic stirring, according to the following methodology:

1. Measure the defined volume of oil in a cylinder.
2. Weigh the proportion of adsorbent material required on the analytical balance.
3. Mix the adsorbent with USO in the beaker and place it on the plate with heating and magnetic stirring at the required temperature, at the specified stirring speed and at the established adsorption time.
4. Filter the mixture in the vacuum pump at 58.86 kPa in order to force the purified oil through the Whatman number 202 filter paper and to retain the adsorbent material.
5. Oil absorbance measurement.

Once the purification was finished, the samples selected to measure their absorbance were diluted to adjust the absorbance between 0.15 and 0.80 after a centrifugation in a T32B JANETZKI centrifuge for 5 min.

Refined soybean oil [clean (L)] was used as a dilution medium, the dilution ratio to be used between pretreated oil [dirty (S)] and the refined oil was 1.0:(3.5–5.25); that is to say that L = (3.5–5.25) S. The total volume for the measurement was 5.0 ml.

2.6 Determination of Absorbance

Absorbance measurement was performed using the Rayleigh VIS-723G spectrophotometer. In the measurements, refined soybean oil was selected as a blank or reference and the wavelength used was 421 nm, according to the result of the UV-Vis spectrum (See Fig. 1S, Supplementary information). The chosen wavelength to the measurement was determined from the maximum absorbance of the spectrum performed on a mixture of refined and used soybean oil.

2.7 Kinetic of Adsorption

In order to develop the kinetic experiment, five samples of USO were selected: initial stage without treatment and the corresponding periods of adsorption over the solid sample of zeolite in separated batches during 1, 2, 4 and 5 h. The other conditions remained the same for all the experiments: Temperature of 35 °C, 10% of zeolite to oil, particle size of 0.63 mm, centrifugation time of 5 min, and centrifugation speed of 8600 rev/min. In each experiment the oil was collected by centrifugation and the absorbance was measured as described in 2.6.

2.8 Experimental Design

The purification was carried out in two stages. The first ones involved the study of the influence of Temperature, Particle size and Proportion of zeolite on the amount of impurities, remaining in the oil after treatment. The latter was estimated from the absorbance of the sample, as it is proportional to dissolved colored compounds. A multilevel factorial experimental ($3^2.2^1$) design with three replicates for a total of 72 experimental runs was used. D-optimun was applied to reduce the number of experiments to 40.

The levels studied for each independent variable are shown in Table 1. The quantities were established according to previous experiences as well as by the consulted bibliography (UCM Rect. and Aracil et al., 2001). Adsorption time of 5 h, as suggested by the kinetic result, and a stirring speed of 1000 min^{-1} were established as fixed variables. Statistical processing was performed using the statistical program Sthatgraphics 19 (Statistical Grraphics, Rockville, MD, USA).

Table 1. The levels of the factors in the experimental design.

Variables	Levels
A Temperature (°C)	80
	100
	120
B Particle size (mm)	0.315
	0.630
	1.000
C Proportion of zeolite (w to w, %)	5
	10

In the second stage of purification, the best design conditions from de the first stage were adopted. The proportion of zeolite was reduced to ensure greater fluidity in the process. In this case, 4% and 6% were used.

3 Results and Discussion

3.1 Characterization of Used Soybean Oil

Different oils were used for the study. The measurement of total polar compounds resulted in a value of 17.5–19%. The density test is performed at room temperature for three samples results in a Mean value of 906 (0.004)-909.30 (0.126) kg/m^3 the standard deviation value appears in parentheses. The degree of acidity was reported as a percentage of oleic acid and the acid number as milligram of KOH per gram of oil. The average value of the degree of acidity and acidity index obtained were 2.26–3.5 (0.1) % and 4.49−7 (0.2) mg of KOH per each gram of oil, respectively. In one of the oils the viscosity was also measured, resulting in a value of 100 mPa.s.

None of these properties are within the parameters established for use edible oil, since the acidity index must be a maximum of 0.6 mg of KOH per gram of oil and the degree of acidity must be less than 0.2% as oleic acid. A high acidity index is a consequence of the presence of water from the fried product, which, together with the high temperatures of the frying process, favored the hydrolysis reaction of the triglycerides, thus increasing the amount of free fatty acids.

The content of polar compounds indicates that the oil is used, but is still suitable because it is less than 25% (NTC 5225, 2013), according to the aforementioned international standards. It should be noted that the obtaining measured density when comparing the density of this used oil with measurements made under the same conditions for unused soybean oil (907.7 kg/m^3), the difference is of the order of 0.2%.

3.2 Characterization of the Natural Zeolite

The composition of the zeolite, used for the purification of the oil, was determined by conventional chemical analysis. The mineralogical composition, determined by X-ray diffraction (XRD), is shown in Fig. 1. The diffractogram shows the presence of three main phases: clinoptilolite (heulandite): $Na_{1.4}Ca_{0.1}K_{0.3}Mg_{0.05}Al_2O_4SiO_{29}\cdot 6.5H_2O$ (JCPDS No. 13–0304), mordenite: $Ca,Na_2,K_2Al_2Si_{10}O_{24}\cdot 7H_2O$ (JCPDS No. 6–0239), and quartz: SiO_2 (JCPDS No. 2–0458).

The composition, stated by calculations from the XRD result shows that zeolite sample was composed of about 70 to 85% of clinoptilolite-mordenite; generally of the sodium calcium type with Si/Al = 4.8 and minor amounts of quartz.

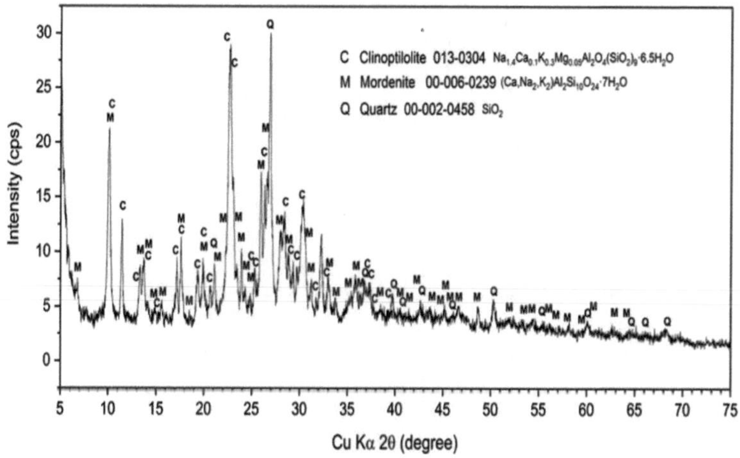

Fig. 1. Diffractogram of the zeolite sample (ZP) used for the purification of used soybean oil.

A similar result is observed when the Raman spectrum of the zeolite is plotted. Figure 2 shows that the Raman spectrum of the studied zeolite sample presents the fundamental bands of the minerals identified by the X-ray diffraction technique: mordenite, heulandite or clinoptilolite and quartz, as compared with reference materials (Lafuente et al., 2015). Such spectra are plotted below of that for the sample. It can be seen that band intensity in the spectrum of the sample is almost the sum of that in the reported reference spectra.

The micrographs of the ZP sample by the Scanning Electron Microscopy (SEM) technique are shown in Fig. 3 (left and right). As can be seen in the image from the left that the zeolite is composed of bundles of crystals of about 1 μm that agglomerate in clusters of up to 15 μm, however, in a larger image (right) it is then observed how the crystals form interlocking frameworks as reported for these types of aqueous-type aluminosilicates (mordenite or ptilolite and heulandite or clinoptilolite) (Milovski and Kononov, 1988).

The structure of zeolites, in particular clinoptilolite and mordenite, is characterized by combinations of Al-O and Si-O tetrahedra that are joined to form large channels and holes. These channels act as molecular sieves and allow the separation of small organic molecules from larger ones. Such is the case with impure soybean oil, where the molecules of degraded components of lower molecular weight and conjugated bonds adsorb light and darken the oil. The laminar morphology of the ZP samples enhances the adsorption of such molecules on the surface of the solid and makes possible the interaction with the crystal structure.

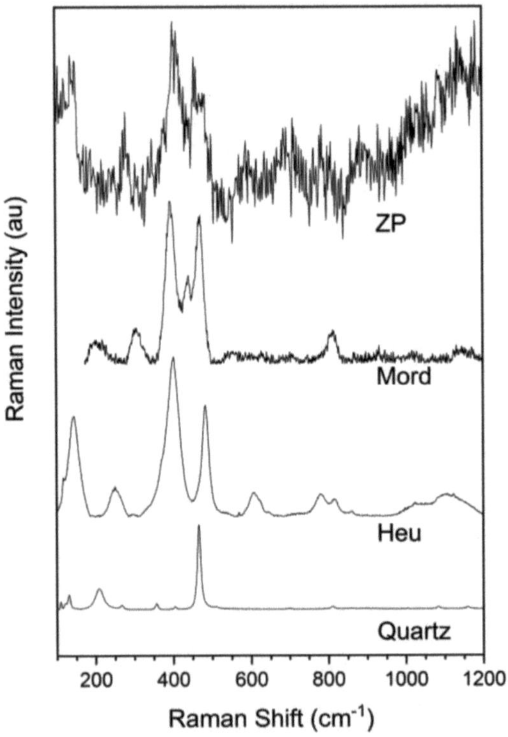

Fig. 2. Raman spectrum of zeolite (ZP) and the corresponding spectra of the main phases present in the sample: mordenite (Mord), heulandite (Heu) and quartz.

Fig. 3. SEM micrographs of the sample ZP, used during purification. Left: 5 kX, right: 20 kX.

3.3 Kinetic Analysis

The variation of absorbance during time of the adsorption process indicated that the concentration of colored compounds was reduced, as is plotted in Fig. 4a. Absorbance exhibits a monotonous and continuous decline which matches closely with an exponential law fit with an R-square value of 0.999. For the experimental conditions selected in the kinetic study, after 4 h of continuous soaking in zeolite, the absorbance of oil is dawn reduced to a 97.3% of the initial value, which indicates that over this time the adsorption achieves near to the equilibrium point. On the other hand as absorbance is related to concentration through the Lambert-Beer law and the property varies proportionally with colored compounds, the actual absorbance (λ) is less of that at initial stage (λ_0) but higher of that in the physical infinity (λ_∞), thus for pseudo-first-order kinetic, a linear relation of $\ln(\lambda\text{-}\lambda_\infty)$ with time arises. The kinetic law of pseudo-first-order rate agrees to the adjusted data of the adsorption experiment, which suggests that this property is dependent on the concentration of colored compounds in the oil (Fig. 4b), even when the other factors studied in the experimental design could vary.

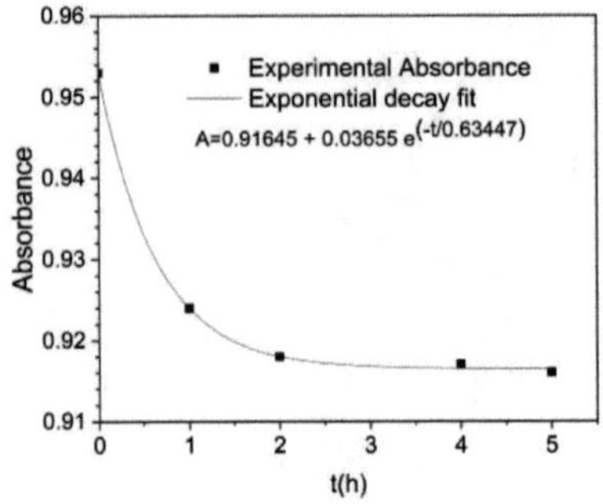

(a)

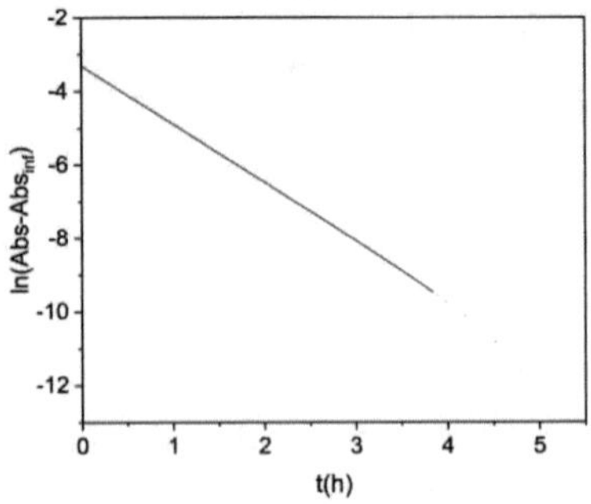

(b)

Fig. 4. Adsorption kinetic of colored compounds in used soybean oil (USO) on the natural zeolite (ZP): (a) Experimental absorbance and exponential decay fit for the selected experiment; (b) Pseudo-first-order plot for adsorption kinetic of data fit on Fig. 4a.

3.4 Statistical Analysis

The experimental results showed a good concordance. The model regression coefficients and the statistical parameters are also presented in Table 2. As can be seen from the table, the coefficient of determination for the response variable reflects that the model explains 92.12% of the variability in impurities. On the other hand, the Durbin-Watson shows that the residuals are independent, with no serial correlation between them, since the P-value is greater than 0.05.

Figure 5 shows the Pareto diagram obtained for the levels of impurities as independent variable. As can be seen, the most significant effect is related to the variable Temperature, while the simpler effect, the proportion of zeolite, is the least significant. The quadratic effect of particle size has a negative influence.

Table 2. Regression coefficients, statistical parameters and optimization for the impurities.

Terms	Coefficients estimates
Independent term	0,1230
A: Temperature	0,0302

(continued)

Table 2. (*continued*)

Terms	Coefficients estimates
B: Particle size	0,0196
C: Amount of adsorbent	0,0035
AA	0,0222
AC	0,0123
BB	−0,0139
R^2 (%)	92,12
Mean absolute error	0,0064
Durbin-Watson	2,6225 (P=0,9229)
Temp optimum (°C)	110 (−0,9573)
Dp optimum (mm)	0.315 (−1)
PZoptimum (%)	10 (1)
Optimum Impurities (minimum)	0,0726

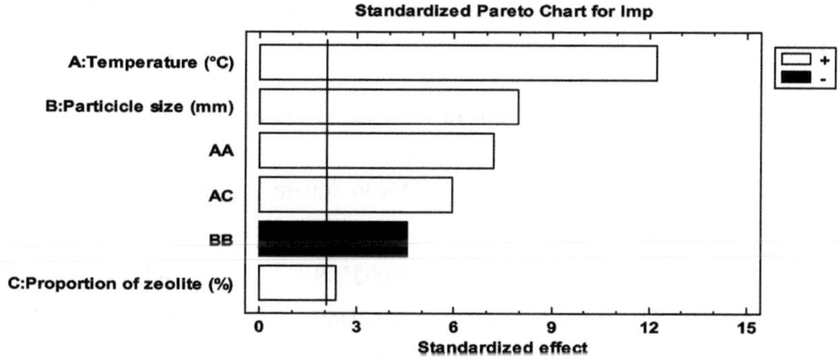

Fig. 5. Standardized Pareto diagram for impurities.

In the analysis of variance (Table 3), all the six effects have a P-value less than 0.05, indicating that they are significantly different from zero with a confidence level of 95.0%, which is in accordance with the goodness-of-fit test. However, the p-value for the block effect is greater than 0.05, resulting in no significance. This is a favorable result because it confirms that the impurity removal process is possible and efficient, regardless of the type of used oil.

As the model complies with the four statistical tests, it adequately represents the behavior of the data.

The effects of the studies variables on impurities is shown in Fig. 6a, b and c. These response surfaces indicate that the lowest impurities value is obtained working at conditions: temperature of 110 °C, particle size between 0.315 mm and with 10% natural zeolite. The minimum value obtained for the response variable was 0.0726. On the surface answer (a) for the first two temperature levels there are practically no differences in the removal of impurities, but as this value increases with the smaller particle size it does decrease. In (b) the decrease is associated with the combination of the low levels of particle size and proportion of zeolite.

Adsorption is an exothermic reaction, so according to the Le-Chatelier principle, forces are strong at low temperatures. In a short time, adsorption will always decrease rising temperature. Regarding the particle size, it is known that adsorption is a surface phenomenon and the magnitude of adsorption should increase with decreasing temperature. With respect to physical adsorption, as it was in the present investigation, it should be taken into account that the "van der Waals", the larger the contact surface, the greater the total amount adsorbed. It is known that, with the same proportion of solids, the surface area is greater with particles of smaller diameter. The results shown in Table 2 are consistent with this analysis.

The second absorption stage confirmed the results obtained in the first experiment: the impurity levels were reduced to 0.093 and 0.084 for 4% and 6% zeolite, respectively, at a fixed temperature and particle size. To verify the quality improvement, a homogeneous transesterification reaction was performed with untreated oil and with oil after the second stage, achieving yields of 85% and 90%, respectively.

Table 3. Analysis of variance for impurities.

Source	Sum of Squares	Df	Mean Square	F-Ratio	P-Value
A:Temp	0,013043	1	0,013043	149,12	0,0000
B:Dp	0,005549	1	0,005549	63,45	0,0000
C:Pz	0,000489	1	0,000489	5,59	0,0247
AA	0,004518	1	0,004518	51,65	0,0000
AC	0,003090	1	0,003090	35,33	0,0000
BB	0,001800	1	0,001800	20,58	0,0001
blocks	0,000028	3	0,000009	0,11	0,9542
Total error	0,002624	30	0,000087		
Total (corr.)	0,033309	39			

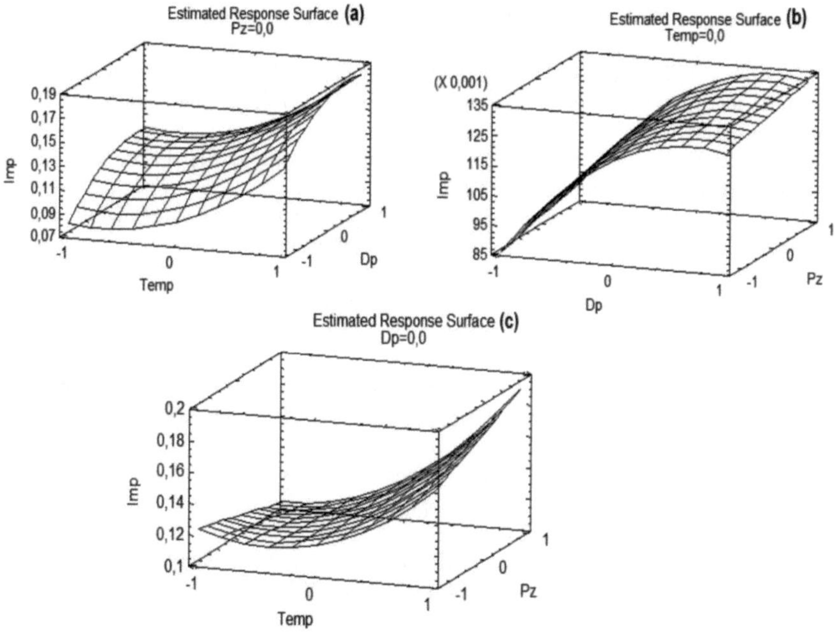

Fig. 6. Response surfaces obtained for the impurities: a) Temperature vs. Particle Size; b) Temperature vs. Proportion of zeolite; c) Particle size vs. Proportion of zeolite.

4 Conclusions

Residual soybean oil from food industry was characterized, obtaining a value of polar compounds of 17.5–19%, a density of 906–909.3 kg/m^3 as well as a degree and acidity index of 2.26–3.5% and 4.49–7 mg/g of KOH-oil, respectively, which corroborating the necessity to discard the oil. The pretreatment process of used soybean oil consists of: residual soybean oil filtration, adsorption stage with zeolite as adsorbent, and phase separation by filtration. From the reported experimental design and statistical study it was shown that lower absorbance values, which is equivalent to a higher adsorption in the purification process, are obtained at 110 °C, with 10% natural zeolite and using particle sizes the 0.315 mm. With the oil after the two stages, a higher yield is obtained in the homogeneous transesterification reaction, which shows that the reduction of impurities is favorable.

Acknowledgements. Authors gratefully acknowledge to the Luces Laboratory at the Materials Research Center, University of Havana for help with Electron Microscopy analysis, to the Materials Department at the Central Criminalistics Laboratory at Havana for help with the Raman Spectroscopy analysis and to Dr. Loreley Morejón at the University of Havana for deliver the XRD experiment.

References

Abdelhameed, R.M., Taha, M., Abdel-Gawad, H., Emam, H.E.: Purification of soybean oil from diazinon insecticide by iron-based metal organic framework: effect of geometrical shape and

simulation study. J. Mol. Struct. **1250**, 131914 (2022). https://doi.org/10.1016/j.molstruc.2021.131914
Ahmad, S.N.S., et al.: Selection of vegetable oils and frying cycles influencing acrylamide formation in the intermittently fried beef nuggets. Foods **10**, 257 (2021). https://doi.org/10.21894/jopr.2016.2802.01
Algieri, C., Drioli, E.: Zeolite membranes: synthesis and applications. Sep. Purif. Technol. (2021). https://doi.org/10.1016/j.seppur.2021.119295
Argyle, M.D., Bartholomew, C.H.: Heterogeneous catalyst deactivation and regeneration: a review. Catalysts **5**(1), 125–269 (2015). https://doi.org/10.3390/catal5010145
Bansal, G., Zhou, W., Barlow, P.J., Joshi, P.S., Lo, H.L., Chung Y.K.: Review of Rapid Tests Available for Measuring the Quality Changes in Frying Oils and Comparison with Standard Methods. Crit. Rev. Food Sci. Nutr. 50(6), 503−514 (2010). https://doi.org/10.1080/10408390802544611.
Boatella-Riera, J., Codony, R.: Recycled cooking oils: assessment of risks for public health; final study. Europ. Parl., Direc. Gen. Res., Direc. A, The STOA Programme. Mr G. Chambers, Ed., Luxembourg (2000)
Cabello, C., Rincón, S., Zepeda, A.: Types of heterogeneous catalysts used for biodiesel production. Afinidad. J. Chem. Eng. Theor. Appl. Chem. 74(577), 51−9 (2017). https://raco.cat/index.php/afinidad/article/view/320756.
Cabezas-Zábala, C.C., Hernández-Torres, B.C., Vargas-Zárate, M.: Fat and oils: Effects on health and global regulation. Rev. Fac. Med. **64**(4), 761−8 (2016). https://doi.org/10.15446/revfacmed.v64n4.53684
Carballo, L.M.: Introduction to Heterogeneous Catalysis (2002). ISBN 9587012070. Bogotá.
Fogler, H.S.: Elements of the Chemical Reaction´s Engineering, fourth ed. Pearson educ. Mexico (2008). ISBN: 978-970-26-1198-1.
Kalbuadi, D.N., Goenadi, D.H., Santi, L.P., Nurtjahja, L.R.: The potential use of natural Clinoptilolite zeolite for crude oil spill removal from sea water. J. Miner. Mater. Characterization Eng. **7**, 446–453 (2019). https://doi.org/10.4236/jmmce.2019.76031
Kesić, Ž, Lukić, I., Zdujić, M., Mojović, L., Skala, D.: Calcium oxide based catalysts for biodiesel production: a review. Chem. Ind. Chem. Eng. Q. **22**(4), 391–408 (2016). https://doi.org/10.2298/CICEQ160203010K
Lafuente, B., Downs, R.T., Yang, H., Stone, N.: The power of databases: the RRUFF project. In: Highlights in Mineralogical Crystallography, T. Armbruster and R.M. Danisi, Eds., Berlin, Germany, W. De Gruyter, pp 1−30 (2015). https://doi.org/10.1515/9783110417104-003
Licata, M.: Frying: changes that occur when frying food. www.zonadiet.com/alimentación/frituras.php. (Accessed 31 Jan 2023).
Loiola, A.R., et al.: Magnetic zeolite composites: classification, synthesis routes, and technological applications. J. Mag. Mag. Mater. **560**, 169651 (2022). https://doi.org/10.1016/j.jmmm.2022.169651
Erickson, M.D., Yevtushenko, D.P., Lu, Z.: Oxidation and thermal degradation of oil during frying: a review of natural antioxidant use. Food Rev. Intl. (2022). https://doi.org/10.1080/87559129.2022.2039689
Mansir, N., Taufiq-Yap, Y.H., Rashid, U., Lokman, I.M.: Investigation of heterogeneous solid acid catalyst performance on low grade feedstocks for biodiesel production: a review. Energy Convers. Manage. **141**, 171–182 (2017). https://doi.org/10.1016/j.enconman.2016.07.037
Milovski, A.V., Kononov, O.V.: Mineralogy. Ed. Mir, Moscow (1988). ISBN 5-03-000612-5.
Mosquera Canchingre, A.F., Cruel Sigüenza, J.A., Tacoronte Morales, J.E., Bernal Villavicencio, C., Canchingre Bone, M.E.: Heterogeneous catalytic system for biofuel generation. Sapienza: Int. J. Interdiscip. Stud. **3**(2), 549−59 (2022). https://doi.org/10.51798/sijis.v3i2.359
NC 660: National Office for Standardization. ISO 660:1996, IDT, Animal and vegetable fats and oils. Determination of acid value and acidity (2021).

NTC 5225: Colombian technical standard. Animal and vegetable fats and oils: determination of polar compound content (2013). https://docplayer.es/56473920-Norma-tecnica-colombiana-5225.html

Rahayu, S., et al.: Environmentally safe technology with the conversion of used cooking oil into soap. J. Phys.: Conf. Ser. ANCOSET **2020**, 012044 (2021). https://doi.org/10.1088/1742-6596/1869/1/012044

Rivera, Y., Gutiérrez, C., Gómez, R., Matute, M., Izaguirre, C.: Quantification of the deterioration of vegetable oils used in frying processes in establishments located in the Libertador Municipality of the State of Mérida. Rev. Cienc. Ing. **35**(3), 157–164 2014). ISSN: 1316-7081. https://www.redalyc.org/articulo.oa?id=507550626005. (Accessed 29 Jan 2023).

Suaterna-Hurtado, A.C.: Deep frying. Perspect. Nutr. Humana **11**(1), 39–53 (2011)

Tacias-Pascacio, V.G., Rosales-Quintero, A., Torrestiana-Sánchez, B.: Evaluation and characterization of residual cooking fats and oils for biodiesel production: a case study. Rev. Int. Contam. Ambient. **32**(3), 303–13 (2016). https://doi.org/10.20937/RICA.2016.32.03.05

Thushari, I., Babel, S.: Comparative study of the environmental impacts of used cooking oil valorization options in Thailand. J. Env. Man. **310**, 114810 (2022). https://doi.org/10.1016/j.jenvman.2022.114810

Tobío-Pérez, I., Díaz, Y., Rodríguez, L., Pohl, S., Lapuerta, M., Piloto-Rodríguez, R.: Biomass-based heterogeneous catalysts for biodiesel production: a comprehensive review. Int. J. Energy Res. **46**(4), 3782–3809 (2021). https://doi.org/10.1002/er.7436

UCM Rect. Aracil, J., Martínez, M., Soriano, R., Guijarrao, M.I., García, T.: inventors, Method for selectively obtaining high value added products from used frying oils. ES Patent No WO2001066679A2. Madrid, España (2001)

Yagüe, M.A.: Study of the use of frying oils in food establishments for prepared meals. Ballaterra, Barcelona. España: Universidad Autónoma de Barcelona (UAB) (2003)

Recycling of Renewable Technologies in Cuba: Photovoltaic

Joel Morales Salas(✉), Elier Cañete Bent, and Carla Viamonte Fernández

Study Center of Renewable Energy Technologies, Technological University of Havana, 114 Street, 11901 Havana, Cuba

jmorales@mecanica.cujae.edu.cu

Abstract. This paper is based on the recycling of renewable technologies in Cuba, specifically photovoltaic solar systems, a topic that is truly of utmost importance due to the awareness that humanity has taken with respect to the search for clean energy and care for the environment. The recycling of this technology emerges as an essential component in the transition towards a sustainable and circular economy. First, the available recycling methods are examined in detail, from chemical and mechanical processes to emerging technologies such as pyrolysis and material recovery. Both their effectiveness in the recovery of valuable materials and their environmental impact are evaluated, including considerations on CO_2 emissions and toxic waste. In addition, the economic viability of solar panel recycling is analyzed, considering the costs associated with collection, transportation and processing, as well as the potential economic benefits derived from the recovery of materials such as silicon, aluminum and glass. In addition, the role of solar panel recycling in the context of the circular economy is addressed, highlighting its contribution to the reduction of electronic waste and the conservation of natural resources.

Keywords: Recycling · sustainable · photovoltaic panels · viability

1 Introduction

Today, 95.1% of Cuba's electricity generation matrix is made up of fossil fuel consumption. The remaining percent is covered by renewable sources [1].

Photovoltaic energy, like the rest of renewable energies, is today a global need to continue developing them. In our case, Cuba, the need is even greater because we are a blockaded country, our fuel supplies are persecuted and we are a tropical archipelago where the sun abounds and we have to take advantage of this natural potential that we have.

According to specialists from the Ministry of Energy and Mines (MINEM), in 2020, until October 2022, 687,000 MWh were generated in the country from renewable energy sources, which represented only 4.9% (3% sugarcane and non-sugarcane biomass, 0.8% solar, 0.5% wind, 0.6% hydropower). This is the equivalent of what 178,000 tons of fossil fuel would have generated, which have been saved by the use of renewable energy.

A. Santos Fuentefria et al. (Eds.): JIFREEEMA 2025, LNNS 1933, pp. 48–57, 2026.
https://doi.org/10.1007/978-3-032-23028-7_5

By 2030, Cuba aims to develop its clean generation program to 29% of its total generation from renewable sources in the energy matrix. These aspirations are more ambitious than those raised in previous years, since there was talk of 24%. To this end, it is proposing the increase of 19 bioelectric power plants fueled by wood and/or sugar cane waste (755 MW), 13 wind farms (633 MW), solar photovoltaic energy (700 MW) and 74 small hydroelectric plants for a total of 2,144 MW of renewable energy.

The savings can be calculated from the cost of a ton of fossil fuel at that time (today, a ton of diesel costs around 500 dollars). 38% of the fossil fuel currently used in generation is imported. With renewable sources, 7,000 GWh would be achieved and approximately 1,800,000 tons of fossil fuel would be saved. Cuba would stop emitting more than six million tons of CO_2 annually. With the call to accelerate this program, mainly in the photovoltaic solar energy sector, progress is being made very quickly because specialists, engineers and technicians from Cuban Electric Company (UNE) have achieved a high level of training and are already designing, building, installing and operating photovoltaic systems with ease. This is a part of the program to which the greatest efforts are currently being dedicated in order to continue increasing the preparation of projects. [1, 2].

But once by 2030 many of the panels already installed will not be in their useful life range, which is why measures need to be taken regarding the recycling of this technology. [1, 2].

The objective of this work is to obtain the best and most efficient way of recycling photovoltaic solar technology in Cuba through different variants according to the country's new business management policy, always taking into account the most viable, economic option that fits the care and protection of the environment.

2 Materials and Methods

The methods to be used are based on the analysis and synthesis of current trends in the recycling of photovoltaic solar technologies, as well as the parameters and standards involved in the recycling of such systems, emphasizing the recycling processes of photovoltaic technology in the world and taking into account the current conditions of recycling in Cuba, in order to understand the feasibility of introducing this recycling industry for photovoltaic technologies installed in the country. Induction and deduction, based on real situations, gathering information and assessing the research process related to the use of photovoltaic technology and its useful life, as well as an economic analysis of the implementation of such a factory for Cuba.

2.1 Analysis of Recycling of Photovoltaic Technologies in the World

The solar industry has experienced rapid growth in recent decades, with a significant increase in the installation of photovoltaic solar panels worldwide. This growth has led to increased generation of waste in the form of solar panels at the end of their useful life, which underlines the importance of recycling to mitigate the environmental impact of these wastes.

Solar panel recycling is crucial to closing the life cycle of these devices and maximizing the recovery of valuable materials, such as silicon, aluminum and glass. By incorporating these recycled materials into the manufacture of new solar panels, the dependence on natural resources is reduced and the generation of waste is minimized, which enters into the chain of the much-valued circular economy.

Solar panel recycling methods include chemical, mechanical and thermal processes. Each recycling method has its own advantages and disadvantages in terms of efficiency, costs and environmental impact. It is important to carefully evaluate these factors when selecting the most suitable recycling method for a particular context.

Some of the main challenges include the lack of adequate infrastructure for recycling solar panels, the scarcity of specific regulations on solar waste management, and the lack of awareness about the importance of recycling among solar industry players.

2.2 Recycling Process of the Panels

Photovoltaic solar panels are composed of various materials, mainly glass (approximately 75% of the total weight), aluminum (for the frame 8%), and silicon photovoltaic cells (5%) encapsulated in polymers (10%) and protected by a back layer. In addition, they contain small amounts of copper, silver, and other metals such as cadmium and lead in some older panel types. This varied composition requires a comprehensive approach to their recycling, given that each material needs a specific treatment for its recovery and reuse.[3].

The typical lifespan of a solar panel is 25 to 30 years. With the exponential growth of the installation of photovoltaic systems in recent decades, a significant wave of end-of-life panels is expected in the coming years. This presents a considerable environmental challenge, as improper disposal of these panels can lead to the release of toxic substances and the loss of valuable materials. It is therefore crucial to develop and optimize efficient and sustainable recycling methods. [3].

2.3 Current Recycling Techniques

The main goal is to recover materials such as glass and aluminum. Glass can be crushed and reused in the manufacture of new glass products, while aluminum is melted and used again in industry.

Disassembly and Delamination

The first step in recycling solar panels by the mechanical method is the initial disassembly, which involves separating the panel from its installation and disassembling the main components, such as the aluminum frame and electrical cables. This process can be done manually or by automated machinery, separating the Aluminum Frame and removing the Cables and Connections [4].

Delamination is the process of separating the layers of solar panels, which are usually composed of a glass layer, an encapsulant layer (usually EVA), silicon photovoltaic cells, and a protective backing layer. There are several methods for delaminating panels:

- Thermal Methods: Thermal methods involve the use of high temperatures to break down the encapsulant materials and release the silicon cells. [5, 6].
- Mechanical Methods: These are the most commonly used methods and consist of physical processes to separate the solar panel components. These processes include the use of crushers and mills to break down the panels into fragments.[4].

- Chemical Methods: Chemical methods use specific reagents to dissolve the encapsulant materials and separate the silicon cells and metals. These methods may include the use of strong acids or bases to attack adhesives and polymers. [7, 8].

2.4 Inverter Recycling

End-of-life inverters are collected and transported to recycling facilities, manually disassembled to separate metal casings and internal electronic components such as metal housings, printed circuit boards (PCBs), cables, and other electronic materials. Metal parts (such as aluminum casings) are recycled by melting them down to make new metal products. Circuit boards and other electronic components are shredded and processed to recover valuable metals such as gold, silver, copper, and other materials. Plastics and other non-metallic materials are recycled or used as fuel in industrial processes. Hazardous components, such as capacitors containing toxic substances, are managed in accordance with environmental regulations to prevent pollution. Inverter recycling not only helps to recover valuable materials and reduce the need for new resource extraction, but is also crucial to minimizing the environmental impact of e-waste. In addition, it prevents the release of toxic substances into the environment and encourages a more sustainable life cycle for technology products.

3 Results and Discussion

3.1 Economic Analysis of the Factory Assembly

In Cuba, for the recycling of photovoltaic technologies according to the components that make up the system, the most feasible would be to recycle the panels, both those found in large photovoltaic parks and those installed in industries, rural areas, polyclinics, etc., but not the inverters and batteries, since the great mass of this technology is found in the parks that supply the National Grid; parks that do not have batteries installed and the inverters are industrial ranging from 500 kW to 1.5 MW, which gives a number of between 4 to 10 inverters per park; according to MINEN specialists, there are 44 photovoltaic parks, that gives an approximate of between 88 to 440 inverters, not as well as the panels that add up to a figure of 700,000 to date in these parks alone.

In the case of batteries, there is already a system for recycling them through the Recycling Group (GER). In 2023, 2,227.9 tons of batteries of all types were collected, 361.6 tons less than the previous year, which represented $889,000 USD for the country. [9].

3.2 Initial Investment Costs

Based on the above estimates, Tables 1 and 2, the total cost to establish a solar photovoltaic technology recycling plant would be approximately between $550,000 and $3.11 million dollars, depending on the number of panels to be recycled. Depending on the range of costs, between 10,000 and 50,000 panels could be recycled annually. [10].

This work does not analyze the labor force because such indicators depend on the time at which the project is decided to be carried out and the economic conditions of the country, as well as the construction and land, since the Recycling Group is the most suitable to take on such a project and has this infrastructure in place.

Table 1. Initial Investment Costs

Initial Costs	Price (USD)
Feasibility and Design Studies:	$40,000 - $80,000
Land and Construction:	-
Equipment and Machinery:	$500 thousand-$3 million
Licenses and Permits:	$10,000 - $30,000
Total, initial estimate:	**$550 thousand-$3.11 million**

Table 2. Annual Operating Costs

Annual Operating Costs	Price (USD)
Labor:	-
Energy and Utilities:	$25,000-$30,000
Maintenance and Replacement Parts:	$35,000-$70,000
Transport and logistics:	$150,000-$200,000
Total, estimated annual:	**$210,000-$300,000**

3.3 Methodology for Approximate Estimation of Unused Panels

Taking into account the data from MINEN specialists, from 2016 to 2024 there are 700,000 solar photovoltaic panels installed in Cuba, between 200 and 300 MW.

To estimate how many of the 700,000 solar panels installed over 8 years may be unused and ready for recycling, several factors must be considered, such as the degradation rate, the premature failure rate, and the annual failure rate. Using data and methodologies based on industry studies, we can calculate a rough estimate [10, 11].

1. Premature Failure Rate: Panels that fail before their expected lifespan (annual failures).
2. Annual Degradation Rate: Loss of efficiency that may lead to the decision to replace panels that still function but are no longer cost effective.
3. Annual Installation Breakdown: Distribution of installations over the 8 years.

These factors are accompanied by the Expected Lifespan of Solar Panels which is 25–30 years, the Premature Failure Rate which is 0.05%-0.5% per year and the Annual Degradation Rate which is 0.5%-1% per year.[11].

3.4 Additional Considerations

- Local Specific Conditions: Factors such as climate, maintenance and installation quality can affect failure rates.

- Degradation and Replacement: Some panels may be replaced not because of total failure but because of significant loss of efficiency (for example, if their performance falls below an acceptable threshold).[12].

For 700,000 solar panels installed over 8 years, it can be estimated that approximately 35,000 panels could be unused and ready to be recycled, considering an annual failure rate of 0.5%, taking as a reference the quality of the panels, the failure rate and the weather conditions of the country (Cyclones, tropical storms and heavy rains with electrical discharges.

3.5 Recoverable Material

A standard crystalline silicon solar panel weighs about 20 kg. The approximate composition is as follows:

- Glass ~ 70% (14 kg).
- Aluminum (frame): ~ 10% (2 kg).
- Silicon: ~ 3% (0.6 kg).
- Plastic (EVA and backing): ~ 10% (2 kg).
- Copper: ~ 1% (0.2 kg).
- Other materials (including small amounts of silver and tin): ~ 6% (1.2 kg). [13].

Recovery rate

The recovery rate may vary depending on the recycling process used, but we can make an estimate [13].

1. Glass: Recovery*: ~95% of 14 kg per panel
   - Total quantity: 35,000 panels * 14 kg/panel * 0.95 = 469,000 kg
2. Aluminum: Recovery: ~95% of 2 kg per panel
   - Total quantity: 35,000 panels * 2 kg/panel * 0.95 = 66,500 kg
3. Silicon: Recovery: ~85% of 0.6 kg per panel
   - Total quantity: 35,000 panels * 0.6 kg/panel * 0.85 = 17,850 kg
4. Plastic: Recovery: ~50% of 2 kg per panel (this can vary greatly depending on recycling method)
   - Total quantity: 35,000 panels * 2 kg/panel * 0.50 = 35,000 kg
5. Copper: Recovery: ~95% of 0.2 kg per panel
   - Total quantity: 35,000 panels * 0.2 kg/panel * 0.95 = 6,650 kg
6. Other materials (including silver and tin): Recovery: ~90% of 1.2 kg per panel.
   - Total quantity: 35,000 panels * 1.2 kg/panel * 0.90 = 37,800 kg

The recycled materials in the Table 3 can be reintegrated into the production chain, promoting sustainability and reducing the need to extract new natural resources.

Estimated value

To estimate the value of the recoverable materials from the 35,000 disused solar panels on the international market, we need to know the average prices of these materials. Prices may vary, but in the Table 4 are estimates based on recent market prices [14–16].

Other materials (including silver and tin): This category is difficult to estimate due to the mix of precious metals and other components, but a conservative estimate would be $40.00/kg (considering the presence of silver in small quantities).[19].

Table 3. Total recoverable materials in 35,000 panels

Materials	Weight (kg)
Glass	469,000 kg
Aluminum	66,500 kg
Silicon	17,850 kg
Plastic	35,000 kg
Copper	6,650 kg
Other materials	37,800 kg

Table 4. Estimated prices of materials (USD per kg)

Recycled material	Estimated prices of materials
Recycled glass	$0.10/kg
Recycled aluminum	$2.00/kg
Recycled silicon	$10.00/kg
Recycled plastic	$0.50/kg
Recycled copper	$8.00/kg

Total Value Calculation.
Glass: Quantity: 469,000 kg

- Price: $0.10/kg
- Total value: 469,000 kg * $0.10/kg = $46,900
  Aluminum: Quantity: 66,500 kg
- Price: $2.00/kg
- Total value: 66,500 kg * $2.00/kg = $133,000
  Silicon: Quantity: 17,850 kg
- Price: $10.00/kg
- Total value: 17,850 kg * $10.00/kg = $178,500
  Plastic: Quantity: 35,000 kg
- Price: $0.50/kg
- Total value: 35,000 kg * $0.50/kg = $17,500
  Copper: Quantity: 6,650 kg
- Price: $8.00/kg
- Total value: 6,650 kg * $8.00/kg = $53,200
  Other materials (including silver and tin): Quantity: 37,800 kg
- Price: $40.00/kg
- Total value 37,800 kg * $40.00/kg = $1,512,000

In the Table 5 are the estimated total value of recoverable materials.

Table 5. Estimated total value of recoverable materials

Materials	Value (USD)
Glass	$46,900
Aluminum	$133,000
Silicon	$178,500
Plastic	$17,500
Copper	$53,200
Other materials	$1,512,000
Total estimated value	**$1,821,400**

3.6 Proposals for a Photovoltaic Technology Recycling System in Cuba

According to the panorama outlined above and the conditions already created in Cuba by the Recycling Group (GER) in terms of infrastructure, logistics, transportation and relevant permits to carry out recycling actions, it would be feasible to build a section for recycling photovoltaic solar panels. GER linked to specialists from the MINEN, who are in charge of decommissioning all those panels that have reached the end of their useful life or no longer generate enough energy for the tasks for which they were installed.

This action would be in line with current trends, where solar panels are recycled in recycling factories for all types of materials (plastic, aluminum, paper, etc.) and not in specialized factories.

The cost of this section would be between $760,000 and $1 million dollars (the cost of a small factory to recycle 10,000 panels per year), which could be possible through international collaboration, since recycling solar panels contributes to caring for the environment, fighting climate change, and promoting the foundations of the circular economy, aspects that are well followed by international organizations that help projects of this nature, including the Green Climate Fund and Cooperation and Development Aid Programs from the European Union.

From an economic perspective, it is feasible, since once the materials are collected and recycled, the estimated total value of the sale of these materials would be $1,821,400 USD.

From a forward-looking perspective and knowing that true energy independence for Cuba lies in renewable energy sources, this recycling section would be a viable option, since the sun is one of the most powerful energy sources the country has and according to the recent policy on photovoltaic energy generation, the installation of 2000 MW is planned for the year 2028, which would represent between 500,000 and 400,000 high-capacity panels.

Using the failure rate, with a higher demand for quality, efficiency and availability parameters, the average annual cumulative failures would be 0.2%, which would represent 800 panels per year, which would represent about $50,000 USD per year.

4 Conclusions

Today, waste from photovoltaic solar panels is recyclable to a very high degree. It is essential to consider that a solar panel is made up of an aluminum frame, silicon photovoltaic cells, a protective glass panel and internal links made of different materials. All of these elements are highly recyclable.

This is why recycling photovoltaic panels is a well-controlled process. It allows for the restoration of a large number of raw materials, which can easily find a second life, a technology that is increasingly growing and that if its materials are recycled and reused, can promote more accessible economic models, greatly reducing production costs.

In Cuba, solar panels are not recycled at the moment, since they are not included in CIRCULAR No. 16/2020, which regulates the products and prices that are recycled by the Raw Materials Recovery Companies, currently named Recycling Business Group (GER), although some of its materials such as plastic, aluminum and silver are included in this circular. Therefore, the results of this article will allow the recycling of photovoltaic solar panels in Cuba to be seen with greater importance.

Processes for recycling photovoltaic solar technology are summarized, the main findings of the study on the construction of a photovoltaic solar panel recycling plant include; Economic viability, which determines that the construction and operation of the recycling plant are financially viable, considering initial investment costs, operating expenses and income from sales of recycled materials.

The environmental benefits of recycling solar panels are identified, including the reduction of electronic waste and the recovery of valuable materials for reuse in the manufacture of new products. The potentially important role that solar panel recycling plants can play in the sustainable management of electronic waste and in the transition to a circular economy is discussed.

Market opportunities for growth in the solar panel recycling sector are suggested, including expansion of recycling capacity and diversification of services offered.

Potential challenges are identified, such as the need to develop more efficient and sustainable recycling technologies, as well as addressing concerns about the management of hazardous chemicals and materials present in solar panels. Areas for future research are suggested, such as optimizing recycling processes, developing more efficient materials recovery methods, and assessing the full life cycle of solar panels and their environmental impact.

In summary, the study shows that building a photovoltaic solar panel recycling plant in Cuba can be a profitable and sustainable investment both economically and environmentally. Positive implications for the recycling industry are highlighted, and key areas for future research and development in this field are pointed out.

5 Recommendations

1) Learn more about the production flow of photovoltaic technology recycling factories around the world, as well as the most up-to-date machinery and its specific price.
2) Learn more about and quantify the useful life of solar panels installed today in Cuba and the number of disabled panels that are no longer used, not only in large parks, but also in small industries, rural areas, schools, polyclinics, etc.

3) Gain more information about the 2000 MW planned for 2028, the quality of the panels, capacity and future projections of Cuba to increase this power in order to achieve energy independence.

References

1. Martín, D.E.S.: Cuba: new legal norms for the development of renewable sources and the efficient use of energy come into force Cubadebate 2019 (Economy, News): p. 11 (2019)
2. Cubadebate, Energía Solar Fotovoltaica en Cuba: Una opción en desarrollo, economica, segura y eficiente (2023)
3. Cuervaenergia: En qué consiste el reciclaje de placas solares? (2023). Available from: https://cuervaenergia.com
4. Choi, J.K., Fthenakis, V.: Crystalline silicon photovoltaic recycling planning: influence of future market scenarios (2014)
5. Chowdhury, M.S., Rahman, K.S., Jaman, A.: An overview of solar photovoltaic panels' end-of-life material recycling (2020)
6. Paiano, A.: Photovoltaic waste assessment in Italy. Renew. Sustain. Energy Rev. **41**, 99–112 (2015)
7. Preet, S., Smith, S.T.: A comprehensive review on the recycling technology of silicon based photovoltaic solar panels: challenges and future outlook. J. Clean. Prod. **448**, 141661 (2024)
8. Berger, W., et al.: A novel approach for the recycling of thin film photovoltaic modules. Resour. Conserv. Recycl. **54**(10), 711–718 (2010)
9. GER, G.E.d.R.: Resultados de la Gestión Empresarial del 2023, 151 (2023)
10. Domínguez, A.G., R.: Photovoltaic waster Assessment Framework and Management in the United States (2019)
11. Jordan, D.C., Kurtz, S.R.: Photovoltaic degradation rates—an analytical review. Prog. Photovoltaics Res. Appl. **21**(1), 12–29 (2013)
12. GL., D.: PV Module Reliability Scorecard 2021. (2021). Available from: https://www.dnv.com/services/pv-module-reliability-scorecard-2018-97974
13. IRENA: End-of-life management: solar photovoltaic panels. Int. Renew. Energy Agency. Report. (2016)
14. World, S.P.: Material Prices, Silicio Reciclado (2024). Available from: https://www.solarpowerworldonline.com
15. News, P.: Material Prices, Plásticos Reciclados (2024). Available from: https://www.plasticsnews.com
16. Metals, K.: Material Prices, Cobre y Plata Reciclados (2024). Available from: (https://www.kitco.com

Valorization of Sacha Inchi Agro-Industrial Waste into High-Protein Flour

Daniela González Franchi[1,2](✉), Mabel Villanueva Domínguez[1], Yanet Sariego Toledo[2], and Luis Cruz Viera[2]

[1] Sacha Inchi Project, Center for Research On Protein Plants and Bionatural Products, 5th Avenue and 246 st., Playa, Havana 11300, Cuba
daniela.franchi@bionaturasm.cu

[2] Faculty of Chemical Engineering, Technological University of Havana "Jose Antonio Echeverría", CUJAE. 114 st. 11901. Marianao, Havana 19390, Cuba

Abstract. This study aimed to valorize Sacha Inchi cake—a byproduct of oil extraction—through the development of a methodology to produce a gluten free, high-protein flour with enhanced nutritional properties. Owing to its rich protein content, Sacha Inchi cake represents a promising alternative to conventional wheat; however, its inherent saponin content demand the implementation of targeted removal strategies. Two desaponification approaches were assessed: a thermal extrusion treatment, varying barrel temperature (80–90 °C and 100–110 °C) and nozzle diameter (8, 12 and 15 mm); and a hydrothermal treatment, conducted at (30, 45 and 60 °C) for (1, 2 and 3 h). Thermal extrusion failed to reduce saponins below the established safety threshold (0.1%). In contrast, hydrothermal processing at 60 °C for 2 h and 3 h achieved saponin reductions of 98.63% and 99.67%, respectively. Both treatments improved the protein profile of the resultant flours, and all processed samples were confirmed gluten free. These findings demonstrate that hydrothermal treatment is an effective, low energy strategy for transforming Sacha Inchi cake into a high value functional ingredient suitable for novel food formulations and the promotion of circular economy principles in agro industry.

Keywords: Sacha Inchi cake · saponins · protein flour · hydrothermal treatment · extrusion

1 Introduction

The growing demand for substitutes to conventional wheat flour is rooted in both dietary and environmental imperatives, reflecting a broader shift toward healthier and more sustainable food systems. This trend is fueled by the rising prevalence of celiac disease (affecting approximately 1% of the global population) and non-celiac gluten sensitivity—conditions that necessitate the elimination of gluten-containing grains such as wheat, barley, and rye from the diet [1]. Concurrently, the push for sustainable agriculture advocates for the diversification of flour-producing crops, favoring species that are resource-efficient and agroecologically adapted, thereby contributing to resilient food systems and reducing reliance on monocultures [2].

A. Santos Fuentefria et al. (Eds.): JIFREEEMA 2025, LNNS 1933, pp. 58–70, 2026.
https://doi.org/10.1007/978-3-032-23028-7_6

Among the emerging candidates for alternative flour sources is Sacha Inchi (*Plukenetia volubilis* L.) cake, a protein-rich by-product obtained during oil extraction from the seeds [3]. With a protein content ranging between 50% and 60%, it offers considerable potential for integration into the food industry [4]. Nevertheless, its utilization in human nutrition is constrained by the presence of saponins—antinutritional compounds known to interfere with nutrient absorption and induce gastrointestinal discomfort [5].

Consequently, developing strategies to reduce the saponin content in Sacha Inchi cake is essential for improving its nutritional quality and broadening its applicability within the food industry. Thermal and hydrothermal treatments have demonstrated efficacy in reducing saponins [6, 7], although they present challenges such as potential protein degradation and the loss of essential nutrients.

This study aims to produce high-protein flour from Sacha Inchi cake by means of optimizing a process capable of reducing saponin content to below 0.1%. Diverging from prior approaches, the proposed methodology prioritizes protein preservation while eliminating the use of chemical solvents, thus enhancing both technical feasibility and sustainability in food system applications.

Repurposing Sacha Inchi cake for protein flour production not only valorizes an underutilized agricultural by-product but also contributes meaningfully to sustainable agricultural practices and circular economy initiatives. In Cuba, for instance, processing 4.3 kg of Sacha Inchi seeds yields approximately 1 L of oil and 1.7 kg of residual cake, corresponding to average extraction yields of 22% for oil and 39.5% for cake (w/w) [8].

Incorporating this residue for food-grade flour production not only reduces the volume of organic waste sent to landfill but also provides a more economical and sustainable alternative to other gluten-free flours, such as quinoa or almond, which are often associated with higher production and commercialization costs.

2 Materials and Methods

2.1 Experimental Material

Sacha Inchi (*Plukenetia volubilis* L.) cake used in this study was procured as a by-product of the oil extraction process conducted at the Basic Production Unit (UBP) "50 Aniversario de la Victoria de Playa Girón" in Güines, Mayabeque, Cuba. The material was transported in sealed polypropylene bags (30 kg each) to the Sacha Inchi oil extraction facility of the Center for Research on Protein Plants and Bionatural Products (CIPB). Prior to experimentation, the cake was manually sorted to eliminate extraneous matter and stored under refrigeration (4 °C) in the dark to preserve its microbiological stability and nutritional composition.

2.2 Thermal Treatment

The effect of extrusion on saponin depletion and nutritional modification of Sacha Inchi cake was investigated using a twin-screw extruder (IBG Monforts DD 85 G, Germany) under a multilevel factorial design ($3^1 \cdot 2^1$), encompassing two temperature ranges (80–90 °C and 100–110 °C) and three nozzle diameters (8, 12, and 15 mm). This resulted in six treatment conditions, each replicated in duplicate. The cake (1 kg) was continuously fed into the extruder and subjected to a high-temperature short-time (HTST) process, producing pellets that were milled (Retsch SM2000, Germany; 0.5 mm mesh) and sieved to obtain a uniform flour. The processed flour was vacuum-sealed and refrigerated (4 °C) to maintain its physicochemical and microbiological integrity.

2.3 Hydrothermal Treatment

A full factorial 3^2 experimental design was employed to evaluate the influence of hydrothermal treatment on saponin reduction and nutritional composition, considering three temperatures (30, 45, and 60 °C) and three treatment durations (1, 2, and 3 h), yielding nine experimental runs with replication. A total of 1 kg of ground cake (Retsch SM2000, Germany; 0.5 mm mesh) was processed with distilled water (1:5, w/v) in 5 L glass beakers and stirred at 800 rpm using a mechanical stirrer (Heidolph RZR-1, Germany). For the 45 °C and 60 °C conditions, a thermostatically controlled water bath (IKA HB4, Germany) was utilized. Following treatment, the suspensions were filtered, and the moist cake was oven-dried. The resulting flour was vacuum-packaged and stored at 4 °C for subsequent analysis.

2.4 Physicochemical Analyses

Key physicochemical parameters—including crude protein, fat, ash, and moisture content—were determined using Near-Infrared Spectroscopy (FOSS NIRS DS2500). Cake samples were pre-ground (Retsch SM2000, Germany; 0.5 mm mesh) to ensure homogeneity, while flour samples were analyzed directly. Measurements were conducted by placing the sample in a small cup, and spectral data were acquired in the near-infrared region. The equipment was calibrated using validated models for Sacha Inchi matrices, enabling rapid, non-invasive, and accurate quantification through automated chemometric processing.

2.5 Saponin Quantification

The saponin concentration of raw and processed Sacha Inchi cake and flours was quantified using a colorimetric-spectrophotometric method, based on the reaction of diosgenin with vanillin under acidic conditions. Absorbance was recorded at 540 nm using a (Rayleigh Vis-723G, P. R. China) spectrophotometer.

A calibration curve was established from a diosgenin stock solution, with eight aliquots ranging from 0.5 to 4.0 mL diluted to 10 mL with 50% ethanol. Each dilution (500 μL) was mixed with 500 μL of 8% vanillin and 5 mL of 72% sulfuric acid. After vortex mixing (IKA Yellowline TTS2, Germany) and incubation at 60 °C for 10 min, absorbance readings were taken.

For experimental samples, 2 g of finely ground material was extracted with 20 mL of 50% ethanol in centrifuge tubes (Hitachi SCT15B, Japan) (5200 rpm, 20 min). A 500 μL aliquot of the supernatant was subjected to the same reaction (vanillin-sulfuric acid) and measurement conditions. For hydrothermal treatments, 500 μL of the aqueous phase was used directly.

Saponin content ((Csap) mg/g) was calculated using the regression equation.

$$\text{Csap} = 10((\text{Abs} - \text{n})/\text{m}) \tag{1}$$

where: Abs represent the measured absorbance, n and m are the intercept and slope of the calibration curve, respectively. For hydrothermal samples, a mass balance approach was employed to estimate the final concentration based on the initial saponin concentration and the fraction removed via aqueous extraction.

2.6 Statistical Analysis

Experimental data were analyzed using StatGraphics Centurion software to determine the effects of thermal and hydrothermal treatments on the physicochemical characteristics and saponin content of Sacha Inchi cake. For both treatment protocols, experimental designs were applied, and linear regression models were fitted for each response variable. Factor significance was determined via analysis of variance (ANOVA), while model quality was assessed using determination coefficients (R2), standard errors, and the Durbin–Watson statistic to verify the independence of residuals.

For calibration curve development, simple linear regression was performed in StatGraphics Centurion. The fitted model was validated through ANOVA, lack-of-fit tests, R2 values, Durbin–Watson statistics, and significance testing to ensure model reliability and predictive accuracy.

2.7 Quality Control Criteria

To ensure product safety and functional performance, key physicochemical quality parameters were monitored, including moisture, gluten, and saponin content. These indicators are especially relevant to the nutritional acceptability and functionality of the final product.

Moisture. Moisture content was the primary parameter routinely monitored, with a critical threshold of 15% to inhibit microbial proliferation and ensure product stability. Measurements were carried out using Near-Infrared Spectroscopy (NIRS) (FOSS DS2500).

Gluten. Given the potential use of the product in gluten-free formulations, the gluten content was restricted to $\leq$20 mg/kg, according to the Cuban Standard NC (2012) [9].

Saponins. The saponin content was limited to a maximum of 0.1% to avoid bitterness and gastrointestinal discomfort. Quantification was conducted as previously described.

3 Results and Discussion

3.1 Physicochemical Composition of Sacha Inchi Cake and Flours.

The physicochemical properties, determined using Near-Infrared Spectroscopy, of Sacha Inchi cake and the derived flours obtained through thermal and hydrothermal treatments are summarized in Table 1. The main parameters evaluated were protein, moisture, fat, and ash content.

Table 1. Physicochemical composition of Sacha Inchi cake and flours.

	Run	Protein (%)	Moisture (%)	Fat (%)	Ash (%)
	Sacha Inchi Cake	43.24 (0.916)	10.2 (0.819)	35.87 (0.198)	5.15 (0.103)
Thermally treated flour	Run	Protein (%)	Moisture (%)	Fat (%)	Ash (%)
	1	49.43 (1.216)	9.60 (0.393)	7.61 (0.937)	6.84 (0.800)
	2	47.41 (1.740)	9.78 (0.937)	9.16 (0.678)	7.00 (0.639)
	3	45.05 (1.627)	9.71 (0.512)	11.86 (0.793)	7.43 (0.564)
	4	52.11 (1.495)	9.18 (0.243)	8.70 (0.670)	6.64 (0.783)
	5	49.34 (1.692)	9.67 (0.861)	9.61 (0.544)	6.64 (0.374)
	6	47.19 (1.885)	9.95 (0.276)	12.64 (0.972)	6.98 (0.706)
Hydrothermally treated flour	Run	Protein (%)	Moisture (%)	Fat (%)	Ash (%)
	1	43.53 (0.278)	10.72 (0.188)	35.90 (0.348)	5.18 (0.316)
	2	43.80 (0.684)	11.04 (0.201)	35.94 (0.391)	5.08 (0.274)
	3	44.36 (0.912)	11.88 (0.217)	36.01 (0.372)	5.03 (0.246)
	4	44.92 (0.531)	12.63 (0.194)	36.08 (0.218)	4.95 (0.301)
	5	45.48 (1.034)	13.44 (0.221)	36.15 (0.356)	4.90 (0.229)
	6	46.04 (1.763)	14.22 (0.209)	36.22 (0.442)	4.86 (0.182)
	7	46.16 (1.145)	14.38 (0.198)	36.23 (0.404)	4.83 (0.265)
	8	46.78 (1.301)	15.19 (0.223)	36.36 (0.387)	4.78 (0.193)
	9	47.41 (1.152)	15.41 (0.224)	36.53 (0.458)	4.53 (0.138)

(): Standard deviation

Sacha Inchi cake. The press cake exhibited high fat (35.87%) and protein content (43.24%), along with moderate ash (5.15%) and low moisture (10.2%), indicating good stability for storage. This composition confirms the cake's potential as a concentrated plant protein source and its origin from an oil-rich seed.

Thermally treated flour. Table 1 shows that thermal treatment increased protein content compared to the raw cake, with values ranging from 45.05% to 52.11%, depending on extrusion temperature and nozzle diameter. The highest protein value (52.11%) was obtained in run 4 (100–110 °C, 8 mm nozzle). Moisture remained relatively low (9.18%–9.95%), which supports good shelf life. Fat content was lower than in the raw cake, ranging from 7.61% to 12.64%, while ash varied between 6.64% and 7.43%, suggesting a moderate concentration of mineral content under some processing conditions.

Hydrothermally treated flour. As shown in Table 1, hydrothermal treatment also led to a slight variation in protein content (43.53%–47.41%) depending on processing temperature and duration. The highest value (47.41%) was recorded in Run 9 (60 °C, 3 h). Moisture increased significantly (10.72%–15.41%) due to water absorption during the treatment, which may compromise product stability unless followed by adequate drying. Fat content remained similar to that of the raw cake (35.90%–36.53%), while ash content slightly decreased (from 5.18% to 4.53%), likely due to mineral leaching.

3.2 Saponin Quantification

To assess saponin removal, saponin content in the cake and flours was measured by the colorimetric–spectrophotometric vanillin–sulfuric acid method using diosgenin standards.

Calibration curve and model fit. Eight diosgenin standards (0.05–0.40 mg/mL) were analyzed at 540 nm. Linear regression of Absorbance vs Concentration yielded $R2 = 0.992$ ($p < 0.0001$) and Durbin–Watson = 2.18 ($p = 0.65$), indicating no autocorrelation. ANOVA gave F Ratio = 2977.4 ($p < 0.0001$), confirming a highly significant fit. Lack of fit testing was not significant ($p = 0.0603$), supporting the adequacy of the linear model

$$\text{Absorbance} = 0.05368 + 0.66718 \cdot \text{Concentration} \tag{2}$$

Saponin content in cake and flours. The quantification of saponins in raw and processed Sacha Inchi flours provided a basis for assessing the efficacy of both thermal and hydrothermal treatments in reducing these antinutritional compounds. Measurements were carried out using the previously described colorimetric method, calibrated against a diosgenin standard curve.

The untreated cake exhibited a saponin content of 56.111 mg/g, which served as the baseline for evaluating treatment efficacy. In thermally processed samples, as shown in Table 2, saponin contents ranged from 37.666 mg/g (Run 4; 100–110 °C, 8 mm nozzle) to 44.860 mg/g (Run 3; 80–90 °C, 15 mm nozzle), with the highest reduction (67.128%) achieved in Run 4. In contrast, hydrothermal treatments yielded greater variability in residual saponin levels, ranging from 4.407 mg/g down to 0.185 mg/g, as summarized in Table 2.

The most pronounced reductions were observed under the most intensive processing conditions—Run 8 (60 °C, 2 h) and Run 9 (60 °C, 3 h)—achieving 98.63% and 99.67% reductions, respectively, relative to the raw cake. These results highlight the superior potential of hydrothermal processing for effective saponin removal, contingent on rigorous control of temperature and treatment duration.

Table 2. Saponin content in Sacha Inchi flours subjected to thermal and hydrothermal treatments.

	Thermally treated flour		Hydrothermally treated flour	
Run	Saponin content (mg/g)	Saponin content (%)	Saponin content (mg/g)	Saponin content (%)
1	37.760 (0.015)	3.776	4.407 (0.087)	0.441
2	39.499 (0.015)	3.950	3.802 (0.023)	0.380
3	44.860 (0.048)	4.486	3.682 (0.017)	0.368
4	37.666 (0.023)	3.767	2.748 (0.026)	0.275
5	42.447 (0.031)	4.245	2.163 (0.052)	0.216
6	43.821 (0.023)	4.382	1.938 (0.026)	0.194
7	–	–	1.264 (0.015)	0.126
8	–	–	0.769 (0.030)	0.077
9	–	–	0.185 (0.015)	0.018

(): Standard deviation

3.3 Influence of Treatments

Thermal treatment. The effects of nozzle diameter (ND) and temperature (T) on the physicochemical properties of Sacha Inchi flour were evaluated using factorial analysis. Table 3 summarizes the direction and statistical significance of each factor on the measured variables. Significant relationships were observed between processing conditions and flour composition.

The nozzle diameter exhibited a positive influence on fat, moisture, and ash content. As ND increased from 8 mm to 15 mm, fat content rose from 7.61% to 12.64%, while moisture increased from 9.18% to 9.95%. These trends can be attributed to reduced oil loss during extrusion and a less uniform heat distribution in larger mass flows [10]. Ash content also increased slightly (up to 7.43%), suggesting that larger diameters may help preserve minerals by limiting direct thermal exposure.

Protein content exhibited an inverse relationship with nozzle diameter, with the highest value (52.11%) recorded using the smallest nozzle. This pattern is likely driven by extrusion dynamics and the thermal inactivation of antinutritional factors. As these compounds are diminished, the relative availability of nutrients increases, resulting in a higher proportion of bioavailable protein. Additionally, the reduction in fat content associated with smaller nozzle diameters contributed proportionally to the relative increase in protein levels.

Temperature also exerted a positive effect on protein content, particularly within the 100–110 °C range, where increases of up to 9% were observed compared to milder conditions. As with nozzle diameter, this effect is attributed to the thermal degradation of antinutritional compounds, which enhances protein bioavailability. In contrast, temperature had a negative impact on ash content, possibly due to the degradation or volatilization of minerals at elevated temperatures [11]. No significant quadratic effects (AA, BB) were detected, while the interaction term (AB) only influenced moisture positively.

Table 3. Effect of process variables on physicochemical properties during thermal treatment.

	Physicochemical properties			
Parameter	Protein (%)	Moisture (%)	Fat (%)	Ash (%)
A: ND	(-)2.3250	(+)0.2200	(+)2.0475	(+)0.2325
B: T	(+)1.1250	NS	NS	(-)0.1683
AA	NS	NS	NS	NS
AB	NS	(+)0.1650	NS	NS
BB	NS	NS	NS	NS
Statistics				
R^2	99.4681	90.4921	89.8681	89.1430
Standard error	0.2282	0.1029	0.6875	0.1252
Mean absolute error	0.1433	0.0556	0.4933	0.0756
Durbin-Watson (p-value)	2.2306 (P = 0.3631)	1.9318 (P = 0.3123)	2.0135 (P = 0.5069)	2.3386 (P = 0.4200)

(+): directly proportional effect, (–): inversely proportional effect, NS: no significant effects.

Statistical analysis confirmed the robustness of the fitted models, with R2 values exceeding 90% and low standard errors. The Durbin-Watson statistic was close to 2, indicating no autocorrelation of residuals ($p > 0.05$), thereby validating the experimental design.

Regarding residual saponin levels, nozzle diameter was the only statistically significant factor ($p < 0.05$), as illustrated in the Pareto diagram (Fig. 1), and Estimated Response Surface (Fig. 2). There is a direct relationship between nozzle diameter (ND) and saponin content (Csap); in other words, larger diameters lead to higher residual saponin levels. This may be due to lower heat penetration in the core of the extrudate when processed through wider nozzles. The limited temperature range evaluated (80–110 °C) failed to produce a significant impact, in contrast to nozzle diameter, which exerted a more substantial effect due to the large difference between its tested levels.

The model exhibited a strong fit to the data, as indicated by an R2 value of 91.561% and a Durbin-Watson statistic of 2.684, confirming the absence of residual autocorrelation ($p = 0.9412$). The mathematical expression of the model is provided in Eq. 3.

$$\text{Csap} = 41.0089 + 3.3138 \cdot \text{ND} \quad (3)$$

This model indicates a linear increase in residual saponin content with increasing nozzle diameter. Flours processed with a 15 mm nozzle exhibited the highest Csap (44.805 mg/g), while those extruded through an 8 mm nozzle had the lowest (37.671 mg/g). However, none of the tested conditions achieved the target threshold (< 0.1%).

These findings underscore the importance of optimizing extrusion geometry to enhance product quality, particularly with respect to reducing antinutritional compounds such as saponins.

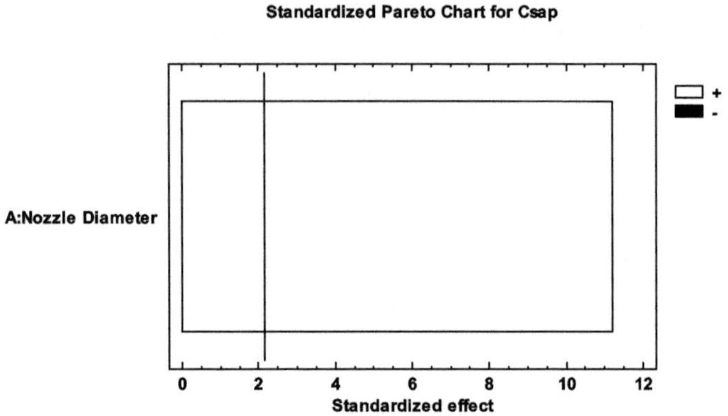

Fig. 1. Pareto chart: factor influence on saponin content during thermal treatment.

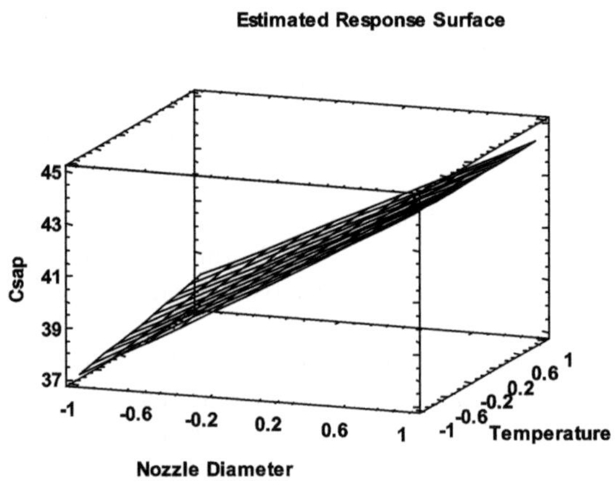

Fig. 2. Estimated Response Surface: factor influence on saponin content during thermal treatment.

Hydrothermal treatment. The effects of processing time (Θ) and temperature (T) on the physicochemical properties of Sacha Inchi flour are summarized in Table 4. Both factors exhibited statistically significant influences on the measured responses, independently and through their interaction.

Higher treatment times (1–3 h) and temperatures (30–60 °C) led to proportionally increased protein content (from 43.53% to 47.41%), likely due to inactivation of saponins and improved protein bioavailability [12]. Moisture content also increased (from 10.72% to 15.41%) as a result of water uptake, indicating a need for post-treatment drying to ensure product stability.

Table 4. Effects of time and temperature on the physicochemical properties of hydrothermally treated flour.

	Physicochemical properties			
Parameter	Protein (%)	Moisture (%)	Fat (%)	Ash (%)
A: Θ	(+)0.5333	(+)0.6300	(+)0.0917	(-)0.0900
B: T	(+)1.4433	(+)1.8900	(+)0.2117	(-)0.1917
AA	NS	NS	NS	NS
AB	(+)0.1050	NS	(+)0.0475	NS
BB	(-)0.1400	NS	NS	NS
Statistics				
R^2	99.8870	98.3377	99.3652	93.4663
Standard error	0.0636	0.2590	0.0205	0.0560
Mean absolute error	0.0307	0.1709	0.0122	0.0364
Durbin-Watson (p-value)	2.9797 (P = 0.6656)	1.5672 (P = 0.1515)	1.6894 (P = 0.1491)	2.0210 (P = 0.3704)

(+): directly proportional effect, (–): inversely proportional effect, NS: no significant effects.

Fat content rose slightly (35.90% to 36.53%), possibly due to thermal disruption of cellular structures and enhanced lipid release. Ash content, on the other hand, decreased from 5.18% to 4.53%, likely due to mineral leaching or volatilization under moist heat conditions [13].

Notably, the interaction term (Θ × T) significantly influenced protein and fat content, with maximal values observed at 60 °C for 3 h. However, the quadratic term for temperature (T2) revealed potential negative effects on protein stability at higher temperatures, suggesting a threshold beyond which structural degradation may occur.

The predictive models exhibited excellent reliability, with R2 exceeding 93% and strong residual independence confirmed by the Durbin-Watson test. These findings validate the robustness of the hydrothermal model.

Saponin content (Csap) decreased markedly under hydrothermal treatment. Both temperature and time were statistically significant ($p < 0.05$), with negative linear effects to Csap, as shown in Fig. 3 and Fig. 4. Longer times and higher temperatures result in lower saponin contents, indicating effective saponin removal under more intense hydrothermal conditions. This behavior is consistent with the literature, as saponins are both heat- and time-sensitive and can be solubilized under controlled moist heat conditions.

The optimal desaponification condition was identified as 60 °C for 3 h, achieving a 99.67% reduction in saponin content. In contrast, milder conditions such as 30 °C for 1 h resulted in only 45% reduction, insufficient for safe food use.

The model's high sensitivity to temperature aligns with prior findings in legumes, where moist heat above 50 °C leads to significant inactivation of saponins. The positive interaction between time and temperature confirms that prolonged exposure to elevated

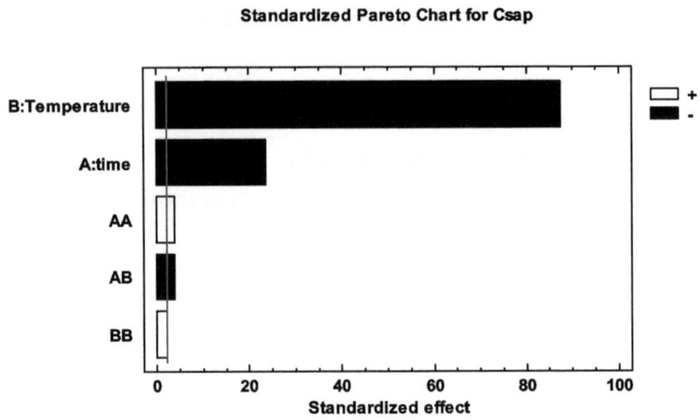

Fig. 3. Pareto chart: factor influence on saponin content during hydrothermal treatment.

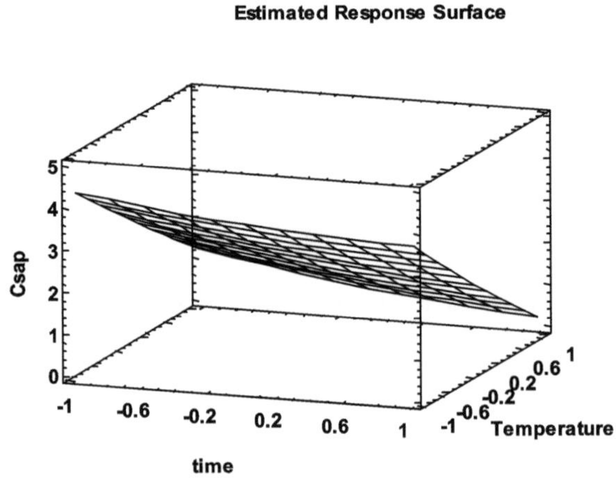

Fig. 4. Estimated Response Surface: factor influence on saponin content during hydrothermal treatment.

temperatures facilitates the effective removal of antinutritional factors while preserving essential nutrients.

The final regression model achieved an R2 of 99.770% and a Durbin-Watson statistic of 3.2453 ($p = 0.9971$), confirming excellent model fit and independence of residuals. The fitted quadratic equation is as follows in Eq. 4.

$$\text{Csap} = 2.1991 - 0.4355 \cdot \theta - 1.6121 \cdot \text{T} + 0.1261 \cdot \theta^2 - 0.0885 \cdot \theta \cdot \text{T} + 0.0683 \cdot \text{T}^2 \quad (4)$$

3.4 Quality Requirements

The evaluation of physicochemical indicators in flours obtained through thermal and hydrothermal processing enabled the assessment of compliance with established quality and safety standards. Three critical parameters were analyzed: moisture, gluten content, and saponins, each of which plays a key role in product shelf life, safety, and nutritional acceptability.

Moisture. All experimental runs from the thermal treatment met the moisture requirement of $\leq 15\%$, which is essential for inhibiting microbial growth and ensuring adequate product stability during storage.

In the hydrothermal treatment, only runs 1 through 7 remained within the acceptable moisture limit. Due to water absorption during hydrothermal processing, additional drying was necessary to adjust final moisture levels and ensure product safety.

Gluten content. All tested samples from both treatments were negative for gluten, remaining below the maximum limit of 20 mg/kg gliadin as specified by Cuban Standard NC (2012). This confirms the suitability of these flours for gluten-free applications and their safety for individuals with celiac disease or gluten intolerance.

Saponin content. The most notable difference between treatments was observed in saponin content. In thermal processing, none of the experimental runs succeeded in reducing saponins below the threshold of 0.1%, indicating that this method, under the tested conditions, was ineffective for saponin removal.

In contrast, hydrothermal treatment proved significantly more effective. Runs 8 and 9, conducted at 60 °C for 2 and 3 h respectively, achieved saponin reductions of 98.63% and 99.67%, thus meeting the safety standard. These findings confirm that hydrothermal processing, when properly controlled, is a viable strategy for achieving saponin elimination in food-grade flour production.

4 Conclusions

The results of this study demonstrate that Sacha Inchi press cake is a nutritionally valuable raw material, with high levels of protein (43.24%) and fat (35.87%), a moderate ash content (5.15%), and low moisture (10.2%), which favors its long-term stability. However, its elevated saponin content (5.611%) poses a limitation for direct food applications duc to thc antinutritional cffccts of thcsc compounds.

Hydrothermal treatment was the most effective method for reducing saponin levels to meet food safety standards. Processing at 60 °C for two and three hours achieved saponin removal rates of 98.63% and 99.67%, respectively, supporting its inclusion as a critical stage in the flour production process for human consumption.

Both processing methods preserved or improved the nutritional properties of the flour, showing notable increases in protein content compared to the raw cake. In thermal treatment, the highest protein level (52.11%) was reached in run 4 (100–110 °C; 8 mm nozzle), while the hydrothermal process reached a maximum of 47.41% in run 9 (60 °C; 3 h). Moreover, all samples tested negative for gluten, confirming the suitability of the resulting flour for gluten-free diets, including those for individuals with celiac disease or gluten sensitivity.

Altogether, these findings support the valorization of Sacha Inchi press cake as a safe and functional ingredient for gluten-free food formulations, contributing to the sustainable management of agro-industrial by-products within a circular economy framework.

References

1. Guandalini, S., Discepolo, V.: *Celiac disease.* Textbook of Pediatric Gastroenterology, Hepatology and Nutrition: A Comprehensive Guide to Practice, pp. 525–548 (2022)
2. Guiñazú Fernández, C.: Harinas sin gluten de origen vegetal para el desarrollo de productos alimentarios. Aplicaciones, propiedades nutricionales y características funcionales. Universitat Politècnica de València: Valencia, España (2020)
3. Vázquez Osorio, D.C., et al.: Desarrollo de galletas empleando harina de Sacha Inchi (Plukenetia Volubilis L.) obtenida de la torta residual. UGCiencia, **23**, 101–113 (2017)
4. Torres Sánchez, E.G., Henández Ledesma, B., Gutiérrez, L.F.: Sacha inchi oil press-cake: physicochemical characteristics, food-related applications and biological activity. Food Rev. Int. **39**(1), 148–159 (2023)
5. Bueno-Borges, L.B., et al.: Sacha inchi seeds from sub-tropical cultivation: effects of roasting on antinutrients, antioxidant capacity and oxidative stability. J. Food Sci. Technol. **55**, 4159–4166 (2018)
6. Mohaddes-Kamranshahi, M., et al.: Evaluation of the saponin green extraction from Ziziphus spina-christi leaves using hydrothermal, microwave and Bain-Marie water bath heating methods. Green Process. Synth. **8**(1), 62–67 (2019)
7. Nikmaram, N., et al.: Effect of extrusion on the anti-nutritional factors of food products: an overview. Food Control **79**, 62–73 (2017)
8. Hernández Pérez, M.C., et al.: Value Chain in the Production of Sacha Inchi Oil (Plukenetia volubilis L.) in Cuba. Int. J. **9**(3), 163–172 (2024)
9. NC: Alimentos para regímenes especiales destinados a personas intolerantes al gluten—Requisitos. In: NC-880, O.N.d. Normalización, Editor. Oficina Nacional de Normalización, Cuba (2012)
10. Carrasco Calle, J.J., Saucedo Guevara, J.M.: Formulación y evaluación de un alimento instantáneo por extrusión a base de maíz mote (Zea mays L.) y trigo (Triticum aestivum) (2019)
11. Alam, M., et al.: Extrusion and extruded products: changes in quality attributes as affected by extrusion process parameters: a review. Crit. Rev. Food Sci. Nutr. **56**(3), 445–473 (2016)
12. Thu, K., et al.: Optimization of hydrothermal treatment on antinutritional factors in selected pigeon pea variety in Myanmar. Roy. Gondwana Res. J. History Sci. Econ. Polit. Soc. Sci. **4**(6), 53–64 (2023)
13. Lagunes-Espinoza, L.C.: Efecto de tratamientos hidrotérmico, remojo y germinación en la composición química de semillas de Lupinus silvestres. Agro Productividad **11**(12) (2018)

Knowledge Transfer and Innovation Management in Regional Universities: A Case Study at the Technical University of Manabí

Roxana J. Panchana-Cedeño[1](✉), Yuniel Bolaño Rodríguez[2], Annette Malleuve Martínez[3], Francis B. Gorozabel-Chata[1], and Daniel Alfonso Robaina[2]

[1] Universidad Técnica de Manabí, Portoviejo, Ecuador
roxana.panchana@utm.edu.ec
[2] Universidad Tecnológica de La Habana "José Antonio Echeverría", La Habana, Cuba
[3] Universidad de Cádiz, Cádiz, Spain

Abstract. The research focuses on the management of technology transfer and innovation at the Technical University of Manabí (UTM) in Ecuador, aiming to analyze the current state of knowledge transfer and propose improvements to this process. A qualitative approach was employed, using the Delphi technique, which involved conducting multiple rounds of questionnaires with a panel of 15 university administrators. This method allowed for the collection and refinement of expert opinions regarding the challenges and opportunities in managing knowledge transfer and technological innovation, as well as identifying potential strategies for implementation at UTM. The most significant findings indicate that UTM faces notable limitations, such as insufficient resources and weak links with the industrial sector, which restrict opportunities for research and innovation. However, there is a demonstrated willingness among administrators to enhance knowledge transfer and foster an innovation-driven culture. In conclusion, it is recommended to establish a Technology Transfer Center that serves as a liaison between the university and local businesses. Additionally, strengthening strategic partnerships with the productive sector and promoting an internal entrepreneurial culture are essential. These actions are crucial for optimizing knowledge transfer management and maximizing the impact of technological innovations at UTM.

Keywords: Management · technology transfer · technological innovation · strategies · collaboration · productive sector

1 Introduction

Technology transfer in higher education institutions (HEIs) has become a strategic pillar for economic and social development, especially in emerging economies where the innovation gap remains significant [1]. According to the Organization for Economic Cooperation and Development [2], while universities in developed countries contribute to over 60% of registered patents annually, in Latin America this figure does not exceed 15%, highlighting a critical disconnect between academic production and its application in the productive sector [3].

A. Santos Fuentefria et al. (Eds.): JIFREEEMA 2025, LNNS 1933, pp. 71–76, 2026.
https://doi.org/10.1007/978-3-032-23028-7_7

In Ecuador, this challenge is even more acute. Recent data from the National Institute of Statistics and Censuses [4] indicate that only 8% of Ecuadorian companies collaborate with universities on R&D projects, one of the lowest rates in the region. This scenario is exacerbated in regional universities such as the Technical University of Manabí (UTM), where structural factors—such as limited funding, a weak innovation culture, and the absence of technology transfer offices (TTO)—hamper the transformation of research into concrete solutions [5].

Recent studies have addressed technology transfer management from multiple perspectives:

- Governance Models: Baldini et al. [6] highlight that efficient technology transfer offices (TTOs) require hybrid structures combining institutional support with operational autonomy. However, in contexts like Ecuador, where 73% of universities lack clear intellectual property policies [7], implementing this approach is challenging.
- University-Industry Collaboration: Ramírez-Correa et al. [8] identified that successful partnerships in Chile and Colombia are based on fiscal incentives for companies, a practice still underdeveloped in Ecuador [9].
- Impact Indicators: Malaver and Vargas [10] propose metrics adapted to Latin American realities, such as the number of prototypes developed, which is more relevant than patents at early stages.

Nevertheless, critical gaps persist:

- Lack of studies on non-capital universities: Approximately 90% of research is concentrated in Quito and Guayaquil [11], ignoring the potential of institutions like UTM in key sectors such as agroindustry and aquaculture.
- Limited evidence on lean models for TTO: While European universities operate with an average budget of USD 2 million annually for transfer activities [2], proven models for institutions with budgets below USD 100,000 are lacking [12].

This research advances knowledge through:

- The first Delphi diagnosis applied to an Ecuadorian regional university, capturing perspectives from key stakeholders (n = 15 administrators).
- A scalable TTO model that integrates: mechanisms for collaborative financing (industry-university-local government), specialized units by economic sector (e.g., Agroindustrial Innovation Center), non-monetary incentive systems (recognition in faculty promotions), and an analysis of public policies, contrasting the new Organic Law of Higher Education [9] with successful cases in Uruguay and Costa Rica [13].

The overall objective of the study is to analyze the current state of knowledge transfer and technological innovation at the Technical University of Manabí (UTM-Ecuador), with the aim of identifying key challenges and proposing strategies to improve management, particularly through the creation of a Technology Transfer Center and strengthening of links with the productive sector.

2 Method

The research adopted a qualitative approach based on the Delphi technique, applied to a panel of 15 university administrators from UTM selected for their experience in university management and innovation. Multiple rounds of structured questionnaires were conducted to:

- Identify consensus on the challenges and opportunities in technology transfer.
- Prioritize improvement strategies using Likert scales and thematic content analysis.
- Validate proposals for a scalable OTT (Office of Technology Transfer) model, adapted to the university's limited resources.

The data were triangulated with secondary sources (reports from SENESCYT, OECD) and comparative case studies (Uruguay, Costa Rica) to contextualize the findings

3 Results

The study revealed significant findings regarding the challenges and opportunities in managing knowledge transfer and technological innovation at the Technical University of Manabí (UTM), structured around three key dimensions:

The study's findings demonstrated a strong consensus among experts (Kendall's coefficient = 0.82, *p* < 0.05) on three critical dimensions for managing technological transfer at UTM: structural challenges, improvement opportunities, and viable strategies for implementation.

3.1 Structural Challenges in Knowledge Transfer

The results revealed critical limitations that hinder the transfer process:

- Insufficient Funding: 93% of managers identified lack of financial resources as the main barrier, with annual budgets below USD 100,000 allocated to innovation (compared to the OECD university average of USD 2 million).
- Low Engagement with the Productive Sector: Only 8% of local companies collaborate with UTM on R&D projects [4], attributed to the absence of fiscal incentives and formalized articulation mechanisms.
- Deficiencies in Institutional Policies: 80% of the panel highlighted the lack of clear regulations for intellectual property protection and the commercialization of research results.

3.2 Opportunities for Innovation

Despite these challenges, enabling factors were identified:

- High Institutional Commitment: 75% of managers expressed willingness to prioritize partnerships with the private sector, especially in strategic sectors such as agribusiness and aquaculture.
- Potential in Untapped Niches: UTM has comparative advantages in areas of regional impact (e.g., prototype development for small industries), which are little addressed by universities in capital cities.

3.3 Evidence-Based Strategic Proposals

Based on the Delphi consensus, specific actions were proposed to optimize technology transfer management. First, a scalable model for a Technology Transfer Office (TTO) was suggested, combining tripartite funding (university-industry-local government) with specialized units for different economic sectors, such as an Agro-industrial Innovation Center. This approach would allow prioritization of resources in high-impact areas, following successful examples documented in the literature [8].

Second, the need to adopt impact metrics tailored to the local context was emphasized. These could include the number of prototypes validated in productive environments and the adoption rate among SMEs, rather than relying solely on traditional indicators like patents. These metrics, aligned with the recommendations of Malaver and Vargas [10], would provide a more realistic assessment of the socio-economic impact of developed innovations.

Finally, the importance of strengthening the culture of innovation within UTM was highlighted through non-monetary incentives. Such measures could include incorporating technology transfer into faculty promotion criteria and establishing a certification for Innovative Researchers. Supported by intrinsic motivation theory, these strategies have proven effective in similar contexts where economic resources are limited, but human capital is abundant.

The findings not only corroborate previous studies on the need for autonomy in TTO [6], but also broaden the discussion by proposing a lean model for universities with limited resources. The success of these strategies will largely depend on the institution's capacity to coordinate internal and external efforts, as well as the adoption of agile management tools to maximize impact with minimal resources. Future research should focus on pilot implementation of these proposals to adapt them to the specific dynamics of UTM and other regional universities in similar contexts.

4 Discussion

This study not only confirms the structural challenges reported in the literature on technology transfer in emerging economies [8, 12] but also advances the conceptualization of an adaptive management model (AMM) specifically designed for regional universities with budget constraints. This model should integrate three interdependent pillars— agile governance, contextualized metrics, and social capital—that respond to the particularities identified at UTM and provide a replicable framework for institutions in similar contexts.

4.1 Agile Governance as the Foundation of the Model

Efforts should be made to establish an OTT with specialized units for each economic sector, challenging the traditional paradigm of centralized offices. This demonstrates that decentralized structures can optimize scarce resources. This approach aligns with This study not only confirms the structural challenges reported in the literature on technology transfer in emerging economies [17] theory of entrepreneurial universities but introduces a critical nuance: the hybridization of roles (researchers, managers-entrepreneurs) within each core, which reduces bureaucracy and accelerates decision-making. Innovation sprints (three-month cycles) and quarterly technical meetings operate as mechanisms for rapid feedback, representing a methodological innovation not previously documented in Latin American studies.

4.2 Contextualized Metrics: Beyond Patents

Replace traditional indicators (e.g., number of patents) with variables reflecting real impact, such as validated prototypes and hours of training provided to SMEs. This transition reflects a paradigmatic shift in evaluating technology transfer, consistent with [10] critiques of measurement systems imported from advanced economies. The data suggest that this approach:

- Increases the perceived utility among local actors (entrepreneurs and faculty), a key factor for sustainability according to diffusion of innovations theory.
- Reduces the gap between research and application by prioritizing scalable solutions over academic publications.

4.3 Social Capital as a Driver of Transfer

The network of mentors composed of alumni entrepreneurs and non-monetary incentives represents a significant theoretical contribution by demonstrating that, in the absence of financial resources, social capital can partially compensate for these limitations. This finding corroborates previous studies in Peru.

4.4 Limitations and Future Directions

While the AMM offers a promising framework, its implementation faces challenges:

- Dependence on institutional leadership: Turnover of authorities in public universities could disrupt processes in the medium term.
- Digital gaps: The success of the demand-driven technology platform requires training SMEs in digital tools, an aspect not addressed in this study.

This study moves beyond diagnosis by proposing an actionable model that:

- Prioritizes adaptability over uncritical replication of international models.
- Institutionalizes frugal innovation, an emerging concept in development literature, but still underutilized in university management.

5 Conclusions

The Technical University of Manabí faces substantial structural limitations in knowledge transfer management, particularly in funding, engagement with the productive sector, and institutional policies. However, there is a clear institutional commitment to improve this process and to capitalize on strategic regional niches.

The AMM not only responds to the needs of UTM but also establishes a methodological precedent for non-capital universities in Latin America, where 68% of institutions face similar challenges [16]. Its value lies in the integration of local evidence, global theoretical frameworks, and agile management techniques—a combination absent in previous models. It is based on agile governance, tripartite funding, and impact metrics tailored to the context. Additionally, it emphasizes the importance of strengthening the culture of innovation through non-monetary incentives and collaboration networks.

References

1. World Intellectual Property Organization (WIPO). Global Innovation Index: Innovation in crisis recovery, p. 2024. WIPO, Geneva (2024)
2. OECD. University-industry collaboration: New evidence and policy options. Paris, OECD Publishing (2023)
3. Fernández-Zubieta, A., Geuna, A., Lawson, C.: Research commercialization in European universities. J. Technol. Transfer. **48**(1), 1–28 (2023)
4. Instituto Nacional de Estadística y Censos (INEC). Encuesta Nacional de Innovación 2023. Quito, INEC (2023)
5. Guzmán, R., López, M., Torres, F.: Barreras a la innovación en universidades ecuatorianas. Rev. Cienc. Adm. **18**(2), 45–60 (2022)
6. Baldini, N., Fini, R., Grimaldi, R.: University technology transfer: framework and trends. Res. Policy **52**(3), 104712 (2023)
7. Secretaría de Educación Superior, Ciencia y Tecnología (SENESCYT). Indicadores de CTeI 2023. Quito, SENESCYT (2023)
8. Ramírez-Correa, P., Rondán-Cataluña, F.J., Arenas-Gaitán, J.: Technology transfer in Latin American universities. Technol. Forecast Soc. Change **186**, 122115 (2023)
9. Asamblea Nacional del Ecuador. Ley Orgánica de Educación Superior (LOES). Registro Oficial **543** (2024)
10. Malaver, F., Vargas, J.: Modelos de transferencia tecnológica en Colombia. Innovar. **32**(84), 19–34 (2022)
11. Vallejo, M., Almeida, J., Cordero, E.: Desigualdades regionales en investigación ecuatoriana. Lat. Am. Res. Rev. **58**(2), 210–228 (2023)
12. Pérez-López, S., Merigó, J.M., Blanco-Mesa, F.: Lean management for university technology transfer. J. Bus. Res. **158**, 113656 (2024)
13. Herrera, L., Sampedro, J.L., García, A.: Políticas de innovación en Centroamérica. Rev. Estud. Soc. **85**(3), 89–104 (2023)
14. Creswell, J.W., Plano Clark, V.L.: Designing and conducting mixed methods research. 4th ed. Los Angeles, SAGE (2023)
15. FAPESP. Annual Report on University Patenting 2023. São Paulo, FAPESP (2023)
16. Science Report, U.N.E.S.C.O.: The race against time for smarter development, p. 2023. UNESCO, Paris (2023)
17. Clark, B.R.: Creating entrepreneurial universities: Organizational pathways of transformation, 3rd edn. Bingley, Emerald (2022)

Reduction of Fuel Consumption in Steam Power Plant by Using Diesel Generator Sets Exhaust

Alberto E. Calvo-González[1(✉)], Luis Enrique Iglesias-Sáez[1], Luis Aballe Infante[1], Jorge E. Arango Gomez[2], Roberto Vizcón Toledo[3], and José R. Rodríguez Bertrán[4]

[1] Faculty of Mechanical Engineering, Renewable Energy Technologies Studies Center, "Jose Antonio Echeverría" Technological University of Havana (Cujae), 19390 Havana, Cuba
aegcalvo@mecanica.cujae.edu.cu

[2] Facultad de Ingeniería, Dpto. de Ingeniería Mecánica y Mecatrónica, Universidad Nacional de Colombia, Bogotá, Colombia

[3] University of Matanzas, Matanzas, Cuba

[4] Technology and Applied Sciences Higher Institute, University of Havana, Havana, Cuba

Abstract. The purpose of this work is to explore the possibility of thermally combining steam reheat power plants with fuel oil-fired diesel generating sets as a combined diesel-steam cycle. These generating sets are electrically connected to the power plant node. The steam unit (mono-block) chosen for this work is 100 MW. Three cases of utilization of waste heat from the fuel oil-fired diesel gen sets in the form of feed water heating repowering are proposed. This way of using the diesel-steam combined cycle as re-powering of the steam power plant under study, modify the classical solution of repowering with gas turbines, when replaced by fuel oil diesel engines. This composition way has not been found in the consulted bibliography. The three variants proposed were calculated and the resulting decrease in fuel consumption was determined for each one. Another original variant, which allows for the utilization of all the possible heat released by the generator sets exhaust gases, also produces reductions and allows for greater operational flexibility, but it must further be studied in detail.

Keywords: Steam power plants · combined cycle · diesel generator sets

1 Introduction

To connect generators sets to the electric grid it is customary to take advantage of the nodes such as substations and steam power plants. But in the case of the node of a steam power plant, if these gen sets are installed as independent units, the advantages that the combination of processes offers, such as the combined diesel-steam cycle, are not taken the possibility of reducing fuel consumption and environmental pollution per unit of energy produced. The convenience of installing diesel-fuel oil generator sets as a combined diesel-steam cycle has been demonstrated in [1] by combining both, steam and diesel. The specific fuel consumption (g/kWh) of the steam cycle was reduced by

A. Santos Fuentefria et al. (Eds.): JIFREEEMA 2025, LNNS 1933, pp. 77–90, 2026.
https://doi.org/10.1007/978-3-032-23028-7_8

15.7% in a 64 MW plant without reheating. The pre-feasibility analysis gave for this solution a Payback Period (RP) of 2 years, 1 month and 16 days. The diesel-steam combined cycle as repowering of the steam power plant modifies the classical solution known of repowering by combined cycle with gas turbines, since these are replaced by diesel engines.

A detailed exposition of repowering and all the ways of approaching it by gas-steam combined cycle was found in literature, it appears in [2], also in [3], but applied to a case study of a 300 MW plant. In a similar way it is done in [4] where they use the **gatecycle™** software. A second law thermodynamic approach to the analysis of the reheat Brayton/Rankine combined power cycle presents in [5]. In [6] the efficiency is increased by adding another gas turbine. The economics of gas turbine repowering is demonstrated in [7]. The closest version to the diesel-steam cycle is the repowering by replacing the regenerative feedwater heaters [8] and in [9] it is in general stated. These results show that the repowering of the feedwater heating system can be an effective strategy to increase the efficiency and power output of conventional steam power plants. Employing the exegertic analysis on a 320 MW power plant [10] reinforces this method of replacing regenerative feedwater heating. The diesel-steam cycle solution has not been found in the consulted bibliography referred to the repowering of steam power plants. The variants used in heat recovery in diesel engines have been scrutinized in the literature and they mainly have been used in small power cycles and organic Rankine cycles [11–14], also using nano-enhanced phase change materials (NEPCM) housed in a plate heat exchanger [15] and solid-state techniques such as thermoelectric generation [16]. A generalization of the methods employed in energy recovery appears in [17], but none to take advantage of it in steam power plants. With the same purpose of decreasing fuel consumption, a new variant of supplier in the heating of feedwater has emerged, in this case with solar radiation. The first article used the term solar repowering, because of its similarity and fulfilled the basic idea [18] and the last published [19–21] also studied the possibility of reducing fuel consumption.

The purpose of this work is to explore the possibility of thermally combining steam power plants with intermediate reheat with fuel oil-fired diesel generator sets as a combined diesel-steam cycle. These generator sets are electrically connected to the power plant node. The power plant under study for this work is a 100MW plant with the following characteristics.

1.1 The Steam Power Plant Under Study Characteristics

The block is composed of a steam generator type BKZ.340/29-M, with a nominal capacity of 340 t/h of superheated steam at a pressure condition of 14.2 MPa (140 ata) and a temperature of 545 °C. The turbine is of the K-100–340/29-M type and consists of three cylinders: a high pressure cylinder (HPC), a medium pressure cylinder (MPC) and a low pressure cylinder (LPC), with a rated electrical power of 100 MW. The superheated steam generated enters the high pressure cylinder with pressure and temperature parameters of 13.2 MPa (130 ata) and 540 °C; once this steam does work in the cylinder, it is extracted (first extraction) to heat the condensate that passes through the high pressure regenerative heater #8; part of the steam can also be sent to the deaerator. The rest of the steam continues to the steam generator passing through the reheater, where it again absorbs heat increasing its energy content to values similar to those of the superheated steam generated. At the outlet of the reheater, the steam enters the medium pressure cylinder, expanding through the 6-stage group that integrates it. In this cylinder there are 6 steam bleedings (2nd, 3rd, 4th, 5th, 6th and 7th) to the high pressure regenerative heater #7, the deaerator, and the low pressure regenerative heaters #5, #4, #3 and #2. The low pressure cylinder has a double flow structure in two opposite directions, where in each direction the 8th bleeding goes to the regenerative heater 1, after which the exhaust steam is directed to the condenser. All of which can be seen in Fig. 1, which shows the Heat Balance Diagram of the block (also named unit) and the data for each point of the HBD are shown in Table 1. The steam turbine of the block chosen for the study is the K-100-130-3600, it is designed to operate with superheated steam at a temperature and pressure at the inlet of the valve of 540 ºC and 13.2 MPa (130 ata). It rotates at an operating speed of 3600 rpm, whose nominal capacity of electrical generation is 100MW.

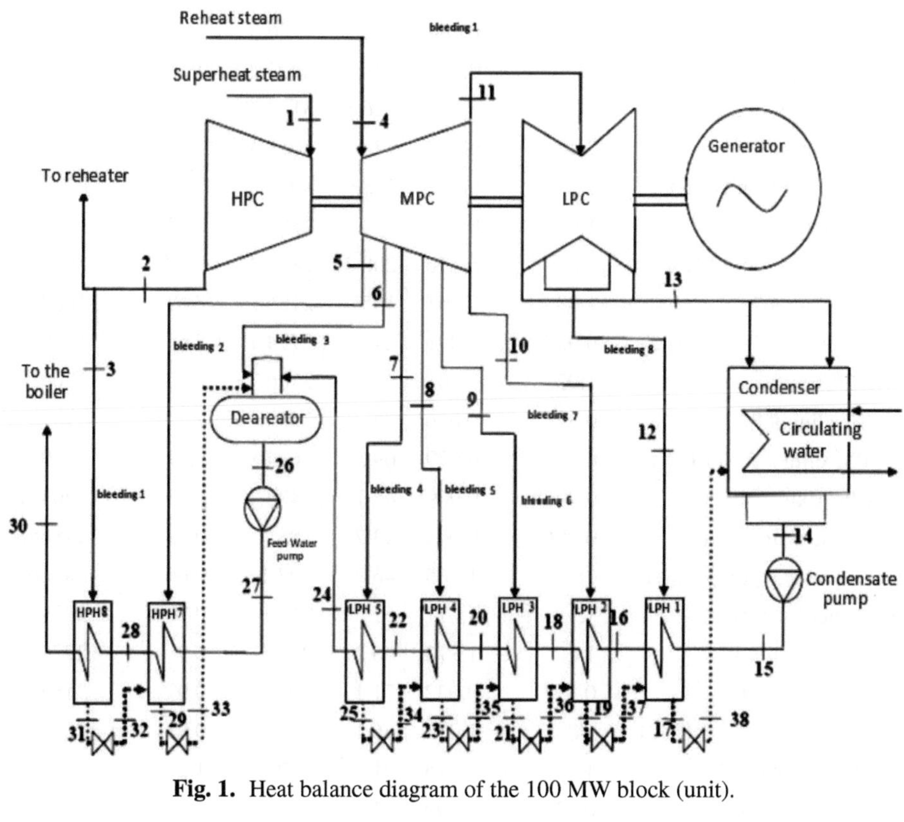

Fig. 1. Heat balance diagram of the 100 MW block (unit).

Table 1. Values of each point of the Heat balance Diagram under study 100MW unit.

MTD 100 MW Reference points	Ata Kg/cm2	MPa	T °C	H (kJ/kg)	S (kJ/kg°C)	Flow (kg/h)
1.Boiler outlet	140	13.73	545	3447.6	6.55	318700
1′.Boiler outlet(tub. Losses)	130	12.75	540	3445.1	6.58	318700
2. HighPress Cylinder outlet	30.63	3.004	349	3112.3	6.73	-
3. Bleeding # 1	29.4	2.884	348	3112.7	6.75	23084
4. Superheated steam	26.83	2.632	545	3561.3	7.42	
4' Superhea steam (tub-leak)	26.83	2.632	540	3550.1	7.4	283222
5. Bleeding # 2	15.74	1.544	463	3392	7.44	17606
5'. Bleeding# 2 in HP7	14.95	1.466	462	3390.8	7.46	20219
6. Bleeding # 3	9.02	0.884	386	3236.3	7.47	1449
6'. Bleeding#3 in deaerator	9.02	0.884	386	3236.3	7.47	4875
7. Bleeding # 4	5.71	0.560	331	3127.2	7.51	8659
7'. Bleeding # 4 in LPH 5	5.37	0.526	326.8	3119.2	7.52	8659
8. Bleeding # 5	3.44	0.337	273	3013.4	7.54	8339
8'. Bleeding #5 on LPH 4	3.2	0.313	268.9	3005.7	7.56	8339
9. Bleeding # 6	1.955	0.191	213	2897.1	7.57	8946
9'. Bleeding # 6 in LPH 3	1.8	0.176	208.3	2888.4	7.59	8946
10. Bleeding # 7	1.03	0.101	155	2786.3	7.62	10683
10'. Bleeding # 7 on LPH 2	0.937	0.091	148.7	2774.7	7.64	10683
11. MPCExit	1.03	0.101	155	2786.3	7.62	220910
12. Bleeding # 8	0.37	0.036	75	2634.9	7.7	579
12'. Bleeding # 8 on LPH 1	0.3	0.029	69.2	2624.8	7.77	3520
13. LPC Exit	0.0965	0.009	45.1	2482.37	7.85	220331
14. Condensatoroutput	0.0965	0.009	43.7	183	0.62	248663
15. Condensatepump outlet	18	1.765	46.7	197.1	0.66	248663
16. Heater inlet #1	15	1.471	51	214.8	0.72	223983
17. Heater outlet #1	14	1.373	65.3	274.5	0.9	223983
18. Heaterdrain #1	0.327	0.032	71.1	297.6	0.97	31649
19. Mixingchamber 3(LP 2)	14	1.373	66.1	277.8	0.91	255632
20. Heater outlet #2	13	1.275	91	382.1	1.2	255632
21. Heater drain #2	0.92	0.090	97.3	407.7	1.28	28129
22. Heater outlet #3	12	1.177	111.6	468.9	1.44	255632
23. Heater drain #3	1.765	0.173	116.3	488.1	1.49	17446
24. Heater outlet #4	11	1.079	131	551.2	1.64	255632

(continued)

Table 1. (*continued*)

MTD 100 MW Reference points	Ata Kg/cm2	MPa	T °C	H (kJ/kg)	S (kJ/kg°C)	Flow (kg/h)
25. Heater drain #4	3.141	0.308	135.1	568.2	1.69	16998
26.MixingChamber4(inletH5)	11	1.079	131.3	552.5	1.65	272630
27. Heater outlet #5	10	0.981	149.6	630.9	1.84	272630
28. Heater drain #5	5.266	0.516	153.8	648.7	1.88	8659
29. Deaerator outlet	6	0.588	158.1	667.3	1.92	338609
30. Feedpump output	193.7	19.00	160.7	689.8	1.93	325074
31.High Pres Heater 7Drain	7.06	0.692	165.6	698.7	2	43303
32. Heater inlet # 8	192.2	18.852	193.2	830.2	2.24	325074
33. High presHeater8 drain	14.349	1.407	196.2	835.3	2.29	23084
34. BoilerInlet	190.7	18.707	236.6	1024.3	2.64	325074

1.2 Characteristics of the Installed Generator Sets

The characteristics of the installed generator sets, MAN 18V48/60 according to the manufacturer's technical data, have a nominal power of 18.9 MW, producing a mass flow of exhaust gases for this nominal power of about 37 kg/s (133.2 t/h), with a specific fuel consumption of 180 g/kWh, which means that 0.945 kg/s (3.40 t/h) of fuel is consumed. Each engine uses 0.945 kg/s (1 ton/h) of saturated steam at 1.247 MPa (12.24 atm) to heat the tracers, the centrifuges, the fuel oil heater, the buster system, the fuel oil tanks and diesel. In the recovery boiler used to generate the steam to heat the aforementioned components of each generator set, the feed water reaches a temperature of (90–110 ^{0}C). Therefore, there will be a mass flow equivalent to 33.69 kg/s (121.3 t/h) exhaust gases available of each engine for heat exchange in the heaters that are the replacement of the regenerative heaters. The outlet exhaust gas temperature is 360 °C and a dew point below 170°C [1]. These generator sets work with internal combustion engines with high efficiencies, because they have a first recovery of the exhaust gases by compressing the incoming air, which causes the exhaust gases not to have as high temperatures as those produced by a gas turbine.

2 The Methodology to Be Followed

The methodology to be followed is based on the study of the Heat Balance Diagram to calculate and specify the values of its characteristics. The calculation of the HBD is based on proposed in Chapter II and III of in the text published by the university "Proyecto de Plantas Termoeléctricas", CUJAE, slight edition, Calvo, A., 1982.

Initially, from the heat balance, the main steam flow $D_{0,}$ required by the turbine to produce power is determined, related to the other flows D_i of the regenerative bleedings through their fractions, where each fraction α_i which is the quotient between the flow going to the extraction $\alpha_i = D_i/D_0$. The results are shown below in Table 2. Subsequently, the proposed combined cycle substitutions are calculated, finding their characteristics and the decrease in fuel consumption.

Table 2. The MTD calculation results

Variable	Calculated value	
$\alpha_1 = D_1/D_0$	0.095	
$\alpha_2 = D_2/D_0$	0.054	
$\alpha_3 = D_3/D_0$	0.015	
$\alpha_4 = D_4/D_0$	0.029	
$\alpha_5 = D_5/D_0$	0.035	
$\alpha_6 = D_6/D_0$	0.029	
$\alpha_7 = D_7/D_0$	0.034	
$\alpha_8 = D_8/D_0$	0.016	
$W_{indicated}$(INDICATED power of steam)	104.10 MW	
$W_{mechanical}$(shaft power)	101.01 MW	
$W_{electric}$(generator terminals power)	99.98 MW	
W_{LPC} (low pressure cylinder INDICATED)	18.83 MW	
W_{MPC} (medium pressure cylinder INDICATED)	55.81 MW	
W_{HPC} (high pressure cylinder INDICATED)	29.46 MW	
η_M(mechanical efficiency)	0.98	
η_{ge}(electric generator efficiency)	0.98	
η_{em}(electro-mechanical efficiency)	0.96	

In the combined cycle proposals presented below, the exhaust gases from the diesel engines heat the feed water flow in heaters designed for this purpose, replacing the regenerative heaters. Three variants are proposed:

1- The replacement of the bleedings 4, 5, 6, 7 and 8 that feed all the low pressure regenerative heaters except the deaerator.
2- The replacement of the bleedings 1 and 2 feeding the high pressure regenerative heaters.
3- The replacement of the bleedings 4 and 5 that feed two low-pressure regenerative heaters in addition to bleedings 1 and 2, corresponding to the high-pressure regenerative heaters.

2.1 Analysis of Each Variant

Variant 1: Replacement of all the low pressure heaters except the deaerator.
By eliminating the bleedings 4, 5, 6, 7 and 8, the work flows D_0 for the cycle will be modified, it will be lower. The steam bleeding fractions for this modification are

recalculated following the methodology described above. The steam bleeding fractions 1, 2 and 3, remain unchanged: $\alpha_1 = 0.09511969$, $\alpha_2 = 0.054411687$, $\alpha_3 = 0.015306724$. The main steam D_0 required for the turbine, maintaining the same power is calculated by the expression (1) which is the known equation of the turbine power, in this case modified by taking into account the eliminated bleedings in this variant 1:

$$W_{\mathbf{turbine}} = D_0^{45678}[(h_1 - h_2) + (1 - \alpha_1)((h_4 - h_5) + (1 - \alpha_1 - \alpha_2)(h_5 - h_6) + (1 - \alpha_1 - \alpha_2 - \alpha_3)(h_6 - h_{11}) + (1 - \alpha_1 - \alpha_2 - \alpha_3)(h_{11} - h_{13})] \quad (1)$$

where: $D_0{}^{45678}$ is the steam flow consumed by the turbine, $W_{turbine}$ is the turbine power in kW, h means the enthalpies kJ/kg and α_i the fractions of the corresponding bleedings.

Keeping the turbine power at 104102 kW, without considering turbine mechanicals losses and electric losses in the generator, substituting values we obtain the required steam flow is 84 kg/s (302.73 t/h) and the fuel consumption is, B = 6.628 kg/s (23.86 t/h) and the specific fuel consumption (b) of 229.21 g/kWh, for which the efficiency of the block (unit)$\eta_{\mathbf{block}}$ is 36%. The heat required in the new substitute low-pressure heaters will be 28900.30 kW obtained by energy balances in the heat balance diagram for the new conditions.

The mass flow of exhaust gases available from each engine for heat exchange in the new heaters is equivalent to 33.69 kg/s (121.3 t/ h). The amount of waste heat that can be obtained from each engine is equal to the flow of gases times their specific heat, which is determined by their composition, in this case is 0.1082 kJ/kg°C, and by the temperature difference (360 – 170) °C, which is summarized in the expression (2):

$$Q_{\mathbf{Emwh}} = G_{\mathbf{gases}} * cp\,(t_{\mathbf{le}} - t_{\mathbf{eh}}) \quad (2)$$

where: Q_{Emwh} (**E**ach **m**otor **w**aste **h**eat) is the required heat flow to be supplied by the exhaust gases, G_{gases} is the required gas flow and t_{le} is the temperatures at which it leaves the engine and t_{eh} exits the heater. It is 6,925.99 kW. Therefore, 4.17 motors would be required, i.e., 5 motors would be needed to supply the heat flow required by the new heaters.

Variant 2-Replacement of high-pressure heaters.

Determination of the steam flow required to produce 100 MW if the bleedings 1 and 2 are eliminated. Recalculating the rest of the steam bleedings by calculating the MTD, it is obtained that: $\alpha_3 = 0.019197$, $\alpha_4 = 0.0336355$, $\alpha_5 = 0.04092$, $\alpha_6 = 0.034639$, $\alpha_7 = 0.040375$, $\alpha_8 = 0.018892$. Keeping the generation $W_{turbine} = 104102$ kW the steam flow that the turbine has to receive is 79.27 kg/s (285.37 t/h), then the fuel consumption is B = 6.34 (22.83 t/h) and the specific consumption equivalent to 219.32 g/kWh, then the block (the unit) efficiency is $\eta_{block} = 38\%$. The heat required by the new high pressure heaters 7 and 8, $Q_{\mathbf{Hr78}}$ is 26516.21 kW, it was obtained performing an energy balance. Therefore, the flow required of diesel engines needed to satisfy the heat demand is calculated using Eq. (2), clearing up the gas flow remains named as Eq. (3):

$$G_{\mathbf{gases}} = Q_{\mathbf{Hr78}}/cp_{\mathbf{gases}}(t_{le} - t_{eh}) \quad (3)$$

$G_{\mathbf{gases}}$ = 26516.21 kW / 1.082 kJ/kgC (360°C – 170°C) = 128.9 kg/s.

Therefore, the number of motors, knowing that each one will provide 33.69 kg/s, will be 3.83, so 4 motors will be needed to guarantee the heat flow necessary for the substitution of the high pressure bleedings 1 and 2.

Variant 3–Replacement simultaneously of low pressure regenerative heaters 4 and 5, and the high pressure regenerative heaters 7 and 8.
If a total of 4 bleedings are removed, 1, 2, 4 and 5, so the corresponding steam fractions α_1, α_2, α_3, and $\alpha_4$are removed as well. By recalculating the remaining steam fractions, we obtain that: $\alpha_3 = 0.019197$, $\alpha_6 = 0.037480$, $\alpha_7 = 0.04368$, $\alpha_8 = 0.020441$. Therefore, to determine the new turbine work flow D_0in Eq. (1) of the turbine adjusted for this variant 3, named as Eq. (4) would be obtained.

$$W_{\mathbf{turbine}} = D_0^{1234}[(h_1 - h_2) + (h_4 - h_6) + (1 - \alpha_3)(h_6 - h_9) + (1 - \alpha_3 - \alpha_6)(h_9 - h_{10}) + (1 - \alpha_3 - \alpha_6 - \alpha_7)(h_{11} - h_{12}) + (1 - \alpha_3 - \alpha_6 - \alpha_7 - \alpha_8)(h_{12} - h_{13})] \quad (4)$$

Substituting the values we have that: $D_0{}^{1245} = 27676.41$ kg/h = 7.68 kg/s and the corresponding new fuel consumption $B = 6.15$ kg/s (22.14 t/ h). Then, the corresponding specific fuel consumption is 212.70 g/kWh. The efficiency of the block will be 0.39, i.e. 39%.

Heat required for new high pressure heaters and low pressure heaters 4 and 5.
Performing an energy balance in the new high pressure heaters #7 and 8, for this circumstance of having been eliminated the 4 and 5, the amount of heat needed to supply the heater 7 is 140.4 kJ/ kg and in the heater 8 is 194.1 kJ/kg and the heater 5 is 76.9 kJ/kg and the heater 4 is 80.74 kJ/kg. Therefore, the total to be supplied by the exhaust gases is 492 kJ/ kg. Considering the feed water flow as 76,878 kg/s (27.67 t/h) the total heat required is 37834.77 kJ/s. Therefore the flow required of diesel engines needed to satisfy the heat demand is calculated using Eq. (2), clearing up the gas flow was named as Eq. (5): Therefore, the number of motors, knowing that each one will provide 33.69 kg/s, will be 5.46, so 6 motors will be needed to guarantee the necessary heat flow in this case.

2.2 Discussion of Results

The comparative analysis of the results obtained in the three variants studied is shown in Table 3:

Table 3. Summary of the results of the three calculated variants

Variant	Steam unit Efficiency η	Specific consumption in each variant g/kWh	% representing Specific consumption	% of Decrease	No. of motors	Total Power ofset MW
1-LPH	0.36	229.21	**94.5**	**5.5**	5	194.5
2-HPH 7–8	0.38	219.32	**90.8**	**9.2**	4	175.6
3-LPH 5- 4 + HPH 7–8	0.39	212.70	**88**	**11.85**	6	213.4
Original values	0.35	241.29	**100**	**0**	0	**100**

It is evident that all variants decrease the specific fuel consumption, they meet the target set. The percentage of decrease increases with the number of the variant: 5.5, 9.2, and 11.85%. The best option of re-powering is by replacing the high pressure heaters and the low pressure heaters 4 and 5 with a percentage decrease of 11.85, but that means that 6 engines are running producing 6 x 18.9 MW (113.4 MW) generating the whole 213.4 MW. The second variant with 4 engines is the second, also by percentage decrease and the four engines produce 75.6 MW, for a total of 175.6 MW. The first variant is the one with the lowest savings and requires 5 engines and produces a total of 194.5 MW. Depending on the load assigned to the plant (steam turbine + engines) the fuel consumption in the steam (monoblock) power plant can be reduced. But the most important conclusion is that all variants decrease the specific fuel consumption. This is a less expensive repowering because the engines are already installed and the only expense would be the replacement heaters and connecting piping. Continuing the analysis we can appreciate the behavior of the fluids in the new heaters and we can see in the temperature profile of both fluids in Fig. 2.

As shown in Fig. 2, the slope of the heating fluid curve for the high pressure section is lower than in the low pressure section. It is also observed that the exergy losses, although not calculated, are lower in the high pressure exchangers, due to the smaller temperature difference between the hot and cold flows. However, in the variants studied, a large amount of heat is wasted in the exhaust gases; this is easily seen in the calculation of the number of engines required to meet the heat flow demand, which is always a fractional number.

A last alternative to be analyzed to avoid such losses is the use of all the heat contained in the exhaust gases of each engine to heat the feed water flow from the temperature at which it leaves the condenser to the boiler inlet.

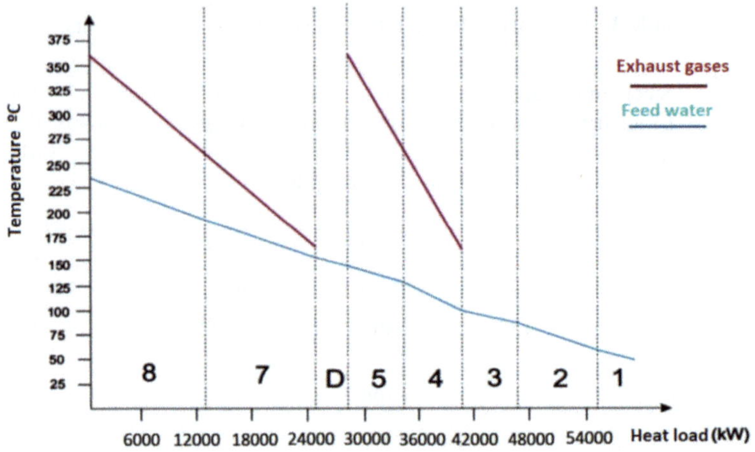

Fig. 2. Temperature profile in heat exchangers

2.3 Utilization of All Heat Released by the Exhaust Gases of Every Generator Sets

It will be determined the consequences, in this case, if the engines are started one at a time until all the heat absorbed by the part of feed water flow, in the low and high pressure systems, is supplied by the exhaust gases. It must be assumed that the block is producing power at the design parameters for the rated load. If the demand is increased, then it would be necessary to start the engines gradually as needed. The heat supplied by each engine were calculated above, Q_{Emwh} (Each motor waste heat) = 6925.99 kW. To determine the mass flow that can be heated by this amount of heat up to the boiler inlet, it is necessary to understand that this feed water flow would be separated before the inlet of the regenerative heater 1and returned to the main line for passage through the deaerator and then heated by the exhaust gases again up to the boiler inlet. It is shown in the following Fig. 3:

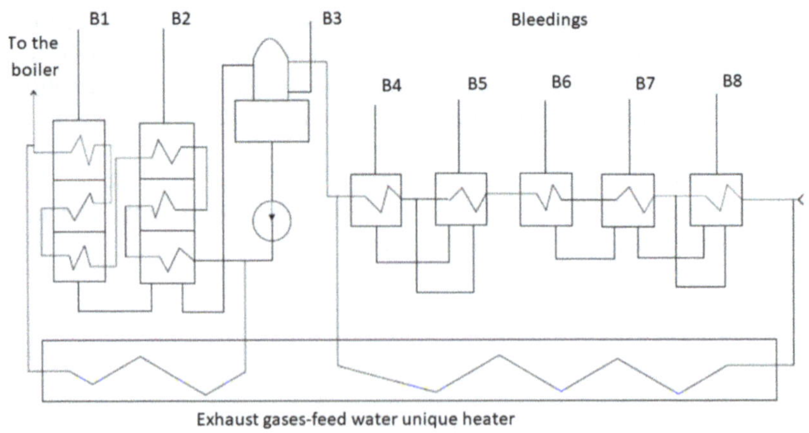

Fig. 3. Simplified heating scheme without replacing heaters

As can be seen in the figure, was simplified the section of the thermal scheme of interest, in addition, the drainage system in the low pressure heaters were modified to simplify the relevant calculations. Once a certain amount of flow feed water is extracted from the main flow for its respective heating by means of the exhaust gases, the bleedings required by the regenerative heaters decrease, then the main flow of steam, D_0, will expand through the turbine to produce 100 MW decreases. This decreases as the steam bleeding for feedwater heating decrease.

As shown in Fig. 3, for a counter flow heater, Eq. (5) can be used, where the heat supplied by the gases is equal to the heat received by the feed water, we have.

$$\mathbf{Q_{deliverbymotors}} = \mathbf{Q_{absorbedbyfeedwater(fw)}} \tag{5}$$

It is known that each motor contributes with $Q_{Emwh} = 6929.$ kW, then the flow of feed water will be able to warm. Clearing up is Eq. (5). Therefore, it is possible to determine the flow of water that can be totally heated by the exhaust gases of an engine up to the boiler inlet point. Substituting in (5) we have (5.1), and by clearance (5.2):

$$N_{\mathbf{motors}} * Q_{\mathbf{Emwh}} = D_{\mathbf{fw}} * (h_{\mathbf{c}} - h_{\mathbf{fw}}) \tag{5.1}$$

$$D_{\mathbf{fw}} = N_{\mathbf{motors}} * Q_{\mathbf{Emwh}}/(h_{\mathbf{c}} - h_{\mathbf{fw}}) \tag{5.2}$$

The decreasing of the water flow passing through the regenerative steam heaters, the efficiency increased by 1% and the specific fuel consumption decreased in 6.82 g/kWh. Since these results were obtained only with exhaust gases generated by one engine, it is clear that the benefits of this variant are undeniable.

The behavior of the block parameters according to the number of engines in operation is shown in the following table (Table 4):

Table 4. Behavior of the block parameters according to the number of engines in operation

number of engines	1	2	3	4	5	6
α_1	0.077	0.069	0.060	0.051	0.042	0.033
α_2	0.044	0.039	0.034	0.028	0.023	0.018
α_3	0.016	0.016	0.017	0.017	0.018	0.018
α_4	0.025	0.022	0.020	0.017	0.014	0.011
α_5	0.025	0.023	0.020	0.017	0.014	0.011
α_6	0.027	0.024	0.021	0.019	0.016	0.012
α_7	0.031	0.028	0.025	0.022	0.018	0.015
α_8	0.014	0.013	0.012	0.010	0.009	0.007
$\eta_{inst.efect}$	0.36	0.36	0.37	0.37	0.38	0.38
D_0 (kg/s)	85.80	84.46	83.16	81.87	80.61	79.37
g/ kWh	234.47	231.16	227.90	224.70	221.55	218.46

(continued)

Table 4. (*continued*)

number of engines	1	2	3	4	5	6
Portion of original value	0.97	0.958	0.94	0.93	0.918	0.905
Savings	0.03	0.042	0.06	0.07	0.082	0.095
Savings	3	4,2	6	7	8,2	9,5

As is evident, although it is a very beneficial variant for the thermal power plant, better results are obtained by completely eliminating the steam bleedings of the high pressure heaters and the low pressure heaters 5 and 4.

3 Conclusions

It has been demonstrated that combining thermally reheated steam power plants with fuel oil-fired diesel generator sets as a diesel-steam combined cycle reduces fuel consumption. The analysis of the results shows that any of the three variants produces the desired effect of reducing fuel consumption at the plant. The variant that most produces the desired effect is the third, replacing high-pressure heaters 7 and 8 with low-pressure heaters 4 and 5. This requires six engines to be operating. Another original variant, which allows for the utilization of all the possible heat released by the generator, sets exhaust gases, also produces reductions and allows for greater operational flexibility, but it must be studied in detail.

3.1 Recommendations

Performing a pre-feasibility study for each of the variants is also recommended. It is also advisable to calculate the reduction in exergy losses in feed water heaters that use exhaust gases, and to consider their potential reduction by proposing a different design. The ways in which this solution can be implemented must be studied to determine how it can be exploited to achieve the greatest reduction in fuel consumption.

Conflict of Interests. The authors declare that they have no conflict of interest with the realization or publication of this work.

References

1. Calvo-González, A.E.; Vizquerra-Rojo, V.: Diesel-Steam combined cycle as repowering of a thermoelectric power plant: case study, Ingeniería Energética Vol. XXXV, No. 2/ 2014 p.122 - 130, May/August, ISSN 1815−5901 Available in: https://www.redalyc.org/ar, Author, F., Author, S.: Title of a proceedings paper. In: Editor, F., Editor, S. (eds.) CONFERENCE 2016, LNCS, vol. 9999, pp. 1–13. Springer, Heidelberg (2016)
2. Stoll, H.G., Smith, R.W., Tomlinson, L.O.: Performance and economic considerations of repowering steam power plants. GER-3644D. GE Industrial & Power Systems (1996)

3. Harutyunyan, A., Badyda, K., Wołowicz, M.: Analyzing of different repowering methods on the example of 300 MW existing steam cycle power plant using gatecycle™ software. Energy **314**, 134228 (2025). https://doi.org/10.1016/j.energy.2024.134228
4. Harutyunyan, A., Badyda, K., Wołowicz, M.: Analyzing of different repowering methods on the example of 300 MW existing steam cycle power plant using gatecycle™ software. Energy **314**, 1 (2025). https://doi.org/10.1016/j.energy.2024.134228
5. Khaliqa, A., Kaushikb, S.: Second-law based thermodynamic analysis of Brayton/Rankine combined power cycle with reheat. Appl. Energy **78**, 179-197 (2004). https://doi.org/10.4236/sgre.2010.13019. Available at: www.elsevier.com. ISSN 1934-9424
6. Carapellucci, R., Milazzo, A.: Repowering combined cycle power plants by a modified STIG configuration. Energy Convers. Manage. **48**, 1590–1600 (2007). https://doi.org/10.1016/j.enconman.2006.11.024
7. Mehrpanahi, A., Hosseinalipour, S.M., Mobini, K.: Investigation of the effects of repowering options on electricity generation cost on Iran steam power plants. Int. J. Sustain. Energ. **32**(4), 1–15 (2011). https://doi.org/10.1080/14786451.2011.622762
8. Mitrović, D., Ignjatović, M., Stojanović, B., Janevski, J., Škundrić, J.: Energy analysis of repowering steam power plants by feed water heating, facta universitatis. Ser.: Mech. Eng. Vol **20**(1) (2022). https://doi.org/10.22190/FUME200218017M
9. Naeimi, J., Biglari, M., Zirak, Gavzan, I.J.: Enhancing conventional steam power plant performance through feed water heating repowering. Energy Sour. Part A: Recovery Utilization Environ. Effects **46** (2024). https://doi.org/10.1080/15567036.2023.2277364
10. Adeli, J., Niknejadi, M., Toghraie, D.: Full repowering of an existing fossil fuel steam power plant in terms of energy, exergy, and environment for efficiency improvement and sustainable development. Environ. Dev. Sustain. **22**, 5965–5999 (2020). https://doi.org/10.1007/s10668-019-00461-x
11. Rijpkema, J., Erlandsson, O., Andersson, S.B., Munch, K.: Exhaust waste heat recovery from truck engine. Energy **238**, 121698 (2022). https://doi.org/10.1016/j.energy.2021.121698
12. Qu, J., Feng, Y., Zhu, Y., Zhou, S., Zhang, W.: Design and thermodynamic analysis of a combined system including steam Rankine cycle, organic Rankine cycle, and power turbine for marine low-speed diesel engine waste heat recovery. Energy Convers. Manage. **245**(1), 114580 (2021). https://doi.org/10.1016/j.enconman.2021.114580
13. Bari, S., Randhawa, T.: Waste exhaust heat recovery in diesel engine by using optimum design and rankine cycle. SAE Techn. Pap. 2023-01-0944 (2023). https://doi.org/10.4271/2023-01-0944
14. Da, L., Qiang, S., Ke, S., Guodong, Z., Shuzhan, B., Guoxiang, L.: Diesel engine waste heat recovery system comprehensive optimization based on system and heat exchanger simulation. Open Phys. **19**, 331–340 (2021). https://doi.org/10.1515/phys-2021-003910.1515/phys-2021-0039
15. Soliman, S., Zhu, S., Xu, L., Dong, J., Cheng, P.: Efficient waste heat recovery system for diesel engines using nano-enhanced phase change materials. Case Stud. Thermal Eng. **14**, 101390 (2021). https://doi.org/10.1016/j.csite.2021.101390
16. Khayum , N., Anbarasu , S., Murugan, S.: Optimizing waste heat recovery in a dual-fuel diesel engine through thermoelectric. J. Renew. Sustain. Energy **16**(6) (2024). https://doi.org/10.1063/5.0237606
17. Noor, A.M., Puteh, R.C., Martinez-Botas, R., Rajoo, S., Romagnoli, A., Basheer, U.M., et al.: Technologies for Waste Heat Energy Recovery from Internal Combustion Engine: A Review, Research Gate, Conference Paper (2015). https://www.researchgate.net/publication/283732724 https://doi.org/10.13140/RG.2.2.14893.90084
18. Zekiyilmazozoğlu, M., Durmaz, A., Baker, D.: Solar repowering of Soma-A thermal power plant. Energy Convers. Manage. **64**, 232–237 (2012). https://doi.org/10.1016/j.enconman.2012.04.019

19. Alotaibi, S., Alotaibi, F., Ibrahim, O.M.: Solar-assisted steam power plant retrofitted with regenerative system using Parabolic Trough Solar Collectors. Energy Rep. **6**, 124–133 (2020). https://doi.org/10.1016/j.egyr.2019.12.019
20. Almutairi, K., Nazari, M.A., Salem, M., Rashidi, M.M., Padmanaban, S., El Haj Assad, M.: A review on applications of solar energy for preheating in power plants. Alexandria Eng. J. **61**(7), 5283–5294 (2022). https://doi.org/10.1016/j.aej.2021.10.04
21. Bari, S.A., Fuad, M., Labib, K.F., Ehsan, M.M., Khan, Y., Hasan, M.M.: Enhancement of thermal power plant performance through solar-assisted feed water heaters: an innovative repowering approach. Energy Convers. Manage. **22**, 100550 (2024). https://doi.org/10.1016/j.ecmx.2024.100550

Experimental Estimation of the Decrease in Performance of Solar PV Modules Due to Deposited Ambiental Particulate Matter

Alina Roig[1(✉)], Ariel Rodríguez[1], Josué Benavides[2], and Anita Rodríguez[3]

[1] Center for Information Management and Energy Development (CUBAENERGIA), Havana, Cuba
arrassi@cubaenergia.cu
[2] Institute of Materials Science and Technology (IMRE), University of Havana, Havana, Cuba
[3] Institute of Cybernetics, Mathematics, and Physics (ICIMAF), Havana, Cuba

Abstract. Deposited particulate matter on the surfaces of PV modules decreases their electricity production by reducing the transmission of solar energy to the PV cells. PV module efficiency refers to the actual efficiency once installed and takes into account factors such as dirt (deposited ambiental particulate matter), climate, orientation, temperature, etc. The objective of this study was to experimentally estimate the loss of PV module efficiency due to deposited particulate matter. A normalized current-voltage (I-V) curve was done for eight monocrystalline silicon half-cell PV modules, exposed for 8, 13, and 24 months, under dirty conditions and after cleaning. In all cases, deposited particulate matter was collected for subsequent mass determination. The main preliminary results obtained were: linear dependence relationship, with a slope of 4.029 (%/(g/m^2)) and $R^2 = 0.9673$, between the loss of performance and the mass per unit area of deposited particulate matter on the PV module and for an average deposited mass of 0.4 g/m^2 the performance of the PV modules studied is reduced by around 1.6% where, if the value of 19.85% of the manufacturer's efficiency reference is taken as 100%, this means an 8.4% reduction in its generation due to contamination by deposited particulate matter. An increase in maximum power (Pm), short-circuit current (Isc) and open-circuit voltage (Voc) was observed in all cases after cleaning the PV modules.

Keywords: Performance Loss · PV Module · Deposited Particulate Matter

1 Introduction

The use of systems that harness solar energy for electricity generation is widely applied in many countries at different latitudes, and its use is becoming increasingly widespread, both due to technological advances and lower component costs, as well as its benefits and ease of use.

However, the impact of air pollution on photovoltaic generation is a global problem, which is expected to be exacerbated by climate change with rising global temperatures

A. Santos Fuentefria et al. (Eds.): JIFREEEMA 2025, LNNS 1933, pp. 91–102, 2026.
https://doi.org/10.1007/978-3-032-23028-7_9

and subsequent droughts. Air pollution consists of the presence of substances or forms of energy that alter the quality of the atmosphere, posing risks and causing harm or discomfort to people and the ecosystem in general [1]. The deposition of dust particles (deposited particulate matter), part of air pollution (also referred to as dirt in much of the literature), on inclined surfaces is a complex and multidimensional phenomenon. When studying the behavior of PV systems under natural conditions, it is necessary to evaluate variables that directly and indirectly impact the maximum utilization of solar energy, as well as the deposition of pollution on the surface of the system. This depends on the degree of pollution in the city or study area, installation factors, and environmental factors such as rain and wind [2]. The layer of deposited dust particles causes optical losses, reducing the number of photons reaching the solar cells and, consequently, decreasing performance.

Studies have evaluated the decrease in module efficiency under different deposited dust conditions: The PV module in Islamabad region had a dust layer over its surface with a density of 6.388 g/m^2 and its efficiency was reduced by 15.08%. In Bahawalpur region, the dust density was observed to be 10.254 g/m^2 which caused the output power to be slashed by 25.42% [3]; In India, it was found that solar energy production is currently reduced by between 17% and 25% in these regions, with approximately equal contributions from suspension particulate matter and deposited particulate matter on PV surfaces [4]; according to [5], based on a review of 104 bibliographic sources, dust deposited on the surface of polycrystalline and monocrystalline photovoltaic modules over one year without cleaning resulted in losses of maximum power output (Pmax) of between 18% and 78%, maximum output current (Imax), and short-circuit current (Isc) of between 23% and 80%. [6] conducted a study where the percentage reduction in output power due to deposited dust typical of the geographic location of Nigeria ranged from 5.92% to 28.78%. The effect of dust accumulation on the performance of a solar photovoltaic plant in the conditions of the city of Dushanbe shown that at a dust density of 0.427 mg/cm^2, the panel performance decreases to 36% [7].

Recently, a study conducted by the University of Moa on the impact of dust on PV module efficiency revealed that, despite rainfall occurring on 71% of days, which is generally a natural cleansing agent, the effect of residual dust on efficiency was 1.63% for PV modules located in the city of Moa [8]. Another study using monocrystalline and polycrystalline panels at SRMIST (India) revealed that dust accumulation resulted in a higher power loss rate (30.48%) in polycrystalline photovoltaic modules compared to monocrystalline modules (14.1%) [9].

In a scenario where PV solar plays a crucial role in the generation of clean energy, it is essential to understand how deposited air pollution (dust particles or ambiental particulate matter) impacts energy generation from PV systems.

Therefore, the objective of this work is to experimentally determine the decrease in PV module performance due to deposited particulate matter. For this purpose, the CUBAENERGIA PV solar installation, located in the Playa municipality of Havana, is used as a reference. Air pollution is primarily due to emissions from transportation and natural dust from the environment.

2 Materials and Methods

2.1 Study Facility

This facility consists of seven arrays of 10 modules each, which constitutes part of the microgrid installed at CUBAENERGIA since July 2022. Four arrays are located in a fixed southward orientation with variable tilt angles (Fig. 1a) and three are tilted 10 degrees and oriented east-west (Fig. 1b). These PV modules are half-cell monocrystalline silicon.

(a) (b)

Fig. 1. Arrays composed of Amerisolar AS-6M30-HC-335W solar modules, (a) fixed south orientation with variable tilt angles, (b) 10-degree tilt and east-west orientation.

To evaluate the behavior of modules exposed and operating without cleaning for periods longer than six months, eight PV modules were used: four without cleaning for eight months, two for 13 months, and two for 24 months. These PV modules had not been cleaned since their installation and were exposed to heavy rain on different occasions.

2.2 Measurement Procedure

Initially, the current-voltage (I-V) curve was plotted for each PV module under dirty conditions.The PV modules were cleaned with distilled water, which was duly collected for later laboratory determination of the deposited particulate matter (Fig. 2). The I-V curve was then plotted again under clean conditions. To perform the curves under dirty and clean conditions, all PV modules were positioned at a 15-degree south orientation. Only eight PV modules were used for this test because the behavior was similar in all cases. This approach avoids excessive handling of PV modules that were installed more than a year ago, which can facilitate internal cell breakage and therefore lead to accelerated deterioration.

The structural condition of the PV modules was verified using Electroluminescence (EL) and Infrared Thermography (RT) techniques to rule out structural failures, humidity, hot spots, and other effects on the study.

Fig. 2. Jar with module samples (right), jar with clean water (left).

These samples were analyzed in the laboratories of the Cienfuegos Environmental Studies Center (CEAC) to determine, after water evaporation, the amount of solid matter deposited on the PV modules.

2.3 Performance of Electroluminescence and Infrared Thermography Techniques

EL images provide information on microcracks and other defects within the cellular material. The electroluminescence technique allows qualitative analysis of the condition of the photovoltaic module, applied to the characterization of silicon photovoltaic modules under different conditions [10]. This, in turn, helps identify faults and problems present or developing within the photovoltaic module [11]. The EL images were obtained under ambient nighttime conditions. A solar electroluminescence imaging camera (model PCO.1300) was used.

Infrared Thermography (IRT) is a non-destructive inspection method for PV module, providing a rapid, real-time distribution of the modules' characteristic features. The evaluation was carried out under the conditions of [12] IEC 62446-3: global incident radiation greater than 600 W/m^2 at the PV module plane, wind speed less than 28 km/h, clear skies, and clean PV modules. A thermal imaging camera (model Fluke TiS 55) was used.

2.4 The Current-Voltage (I-V) Curve Plotting

The typical system used for I-V curve measurement consists of natural solar radiation or a solar simulator, a module temperature sensor, a calibrated pyranometer or solar cell, and a data storage system to save measurements when the voltage or current across the module is varied with an external electronic load or power source.

Under natural solar radiation conditions, a portable tracer can be used to measure PV module I-V curves. The problem with this type of inspection is that they are performed on-site, outside of standard measurement conditions. Therefore, the collected values must be corrected to allow a comparison of the electrical parameters under OTC (operational test conditions) with those provided by the manufacturer (Data Sheet) under nominal STC conditions. To correct the measured curves for each PV module, the procedure described in the IEC 60891 standard [13] is used under open-air conditions. The curve obtained is subsequently transformed to standard measurement conditions (temperature 25 °C, solar radiation 1000 W/m^2, and AM 1.5G). This procedure allows for the analysis of some of the potential causes of problems these PV modules may present, which in turn affect the performance of PV installations.

A curve tracer was used (model HT400 IU), coupled with a radiation sensor. Based on the PV module's operating temperature and the manufacturer's data, standardized I-V curves are obtained at 1000 W/m^2 and 25° Celsius. A comparison is made of the PV module's behavior before and after cleaning, and even with its initial technical data according to the manufacturer. Since the measurements were made with natural solar radiation, they were carried out in short time intervals to prevent cloud interference during outdoor measurements.

2.5 Data Processing

The characteristic parameters are determined from the I-V curves:

- Short-circuit current (Isc): This is the maximum current that the device will produce under defined radiation and temperature conditions, corresponding to a voltage equal to zero.
- Open-circuit voltage (Voc): This is the maximum voltage of the device under defined radiation and temperature conditions, corresponding to a current equal to zero.
- Maximum power (Pm): This is the maximum power that the device will produce under given radiation and temperature conditions, corresponding to the maximum current-voltage pair.
- Voltage at the maximum power point (Vmp): This is the voltage value for Pm under given radiation and temperature conditions.
- Current at the maximum power point (Imp): This is the current value for Pm under given radiation and temperature conditions.
- Shape Factor (SF), which is the value corresponding to the quotient of the maximum power and the product of the short-circuit current and the open-circuit voltage (Eq. 1). This gives us an idea of the quality of the PV device, with the higher its shape factor being better, closer to 1 or 100% depending on how it is expressed.

$$FF = \frac{Pm}{Voc \cdot Isc} \tag{1}$$

where:

Pm is the Maximum power of the PV device (W)
Voc is the Open-circuit voltage of the PV device (V)
Isc is the Short-circuit current of the PV device (A)

The characterization described in the DataSheet is used as a reference for the comparison. It should be noted that the data in the DataSheet do not necessarily exactly match the values of the PV module in question, as they are average values obtained in the manufacturing batch. Therefore, if the power loss is not excessive, the calculation error may be greater than the degradation itself. The performance (actual operating efficiency) of each PV module is determined according to Eq. 2.

$$\text{Performance} = \frac{Imp \cdot Vmp}{\text{G} \cdot \text{A}} \cdot 100 \tag{2}$$

where:

Imp is the current value for Pm of the PV device under given radiation and temperature conditions (A)

Vmp is the voltage value for Pm of the PV device under given radiation and temperature conditions (V)

G is the incident radiation density (W/m^2)

A is an area of solar module (m^2)

From all sampled values, the particulate matter per unit area was determined, the performance values (actual instantaneous efficiency) were determined for each case, and the impact of deposited particulate matter on the PV module on their performance was analyzed based on a comparative study of the I-V curve.

From the graphic of the measuremed values, the relationship between the decrease in performance and the grams of particulate matter deposited on the PV module per unit area was obtained.

3 Results and Discussion

3.1 Electroluminescence (EL) and Infrared Thermography (RT) Techniques Sample

Based on the implementation of both techniques, it was verified that there were no structural problems that influenced the performance decrease of the evaluated PV modules. Figure 3 shows the images taken of two of the PV modules used in the evaluation. In the case of EL, the image obtained of the PV module before installation is compared with that obtained after 12 months of operation. In the case of RT, a homogeneous operating temperature distribution is observed throughout the PV module.

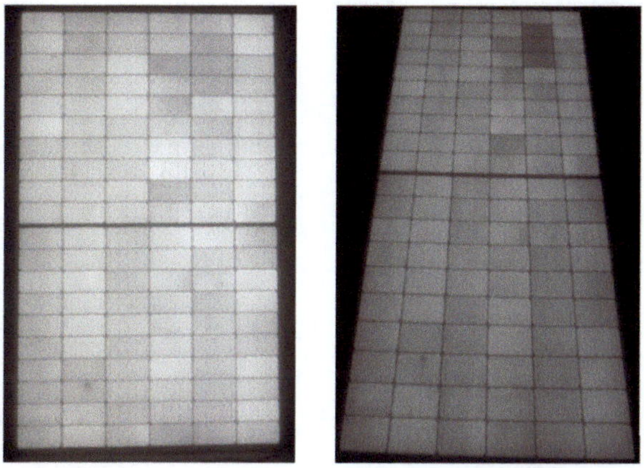

Case 1. EL

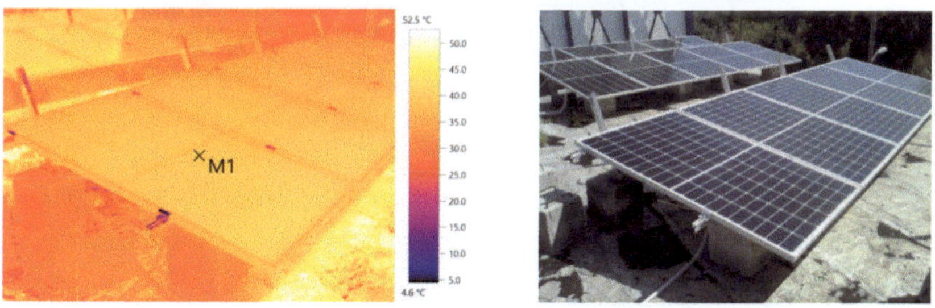

Case 2. RT

Fig. 3. Sample of EL and TR images obtained from the PV modules used in the study.

3.2 The Current-Voltage (I-V) Curve Plotting

Figure 4 shows the behavior of the I-V curve for 3 of the PV modules evaluated before and after cleaning.

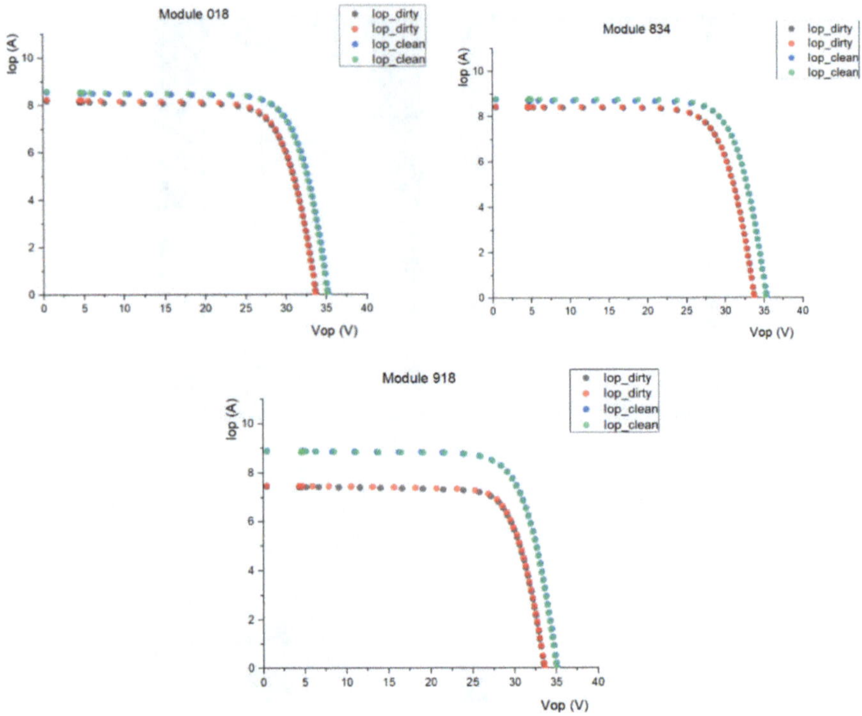

Fig. 4. I-V curves plotted for PV module evaluated before and after cleaning

Figure 4 shows how, after cleaning, both the open-circuit voltage (Voc) and the short-circuit current (Isc) increase for PV modules, suggesting an increase in generation in all cases. Figure 5 shows how, after cleaning, the power vs. voltage curve is always higher, meaning that the maximum power delivered increases in all cases.

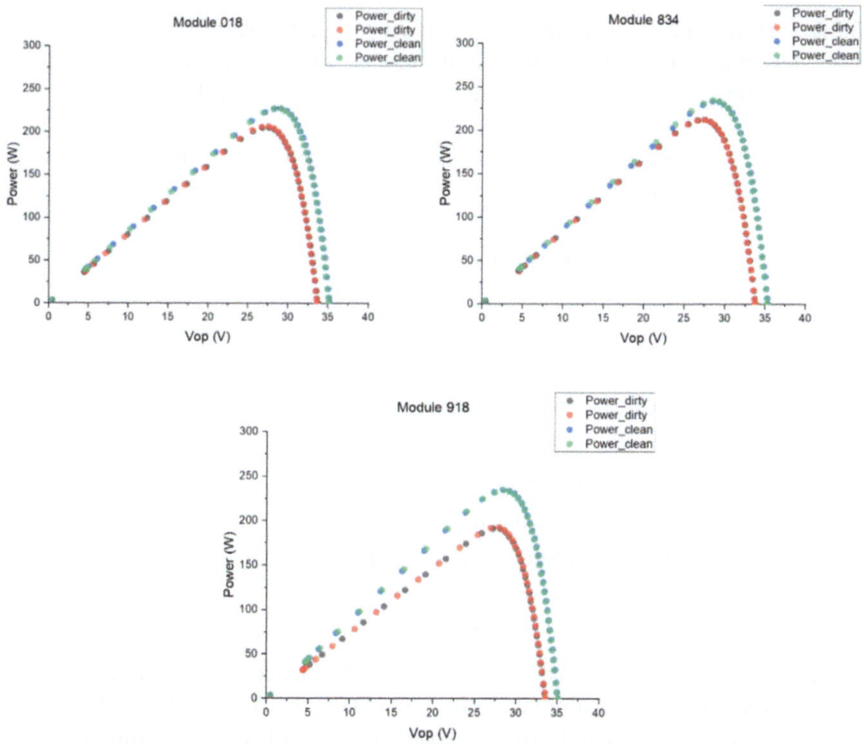

Fig. 5. Power curves plotted for PV module evaluated before and after cleaning

Table 1 shows the average results for PV module characteristic parameters obtained under dirty and clean conditions.

Table 1. Average results obtained for dirty and clean PV modules from the I-V curve plot, module manufacturer's reference data.

Parameters	Data Sheet	Average Dirty	Average Clean
Actual instantaneous efficiency (%)	19.85*	14.98	16.53
Vmpp(V)	34.40	27.96	28.58
Impp(A)	9.74	7.86	8.58
Pm (kW)	335.00	220.03	244.99
Voc(V)	41.00	34.22	35.21
Isc(A)	10.24	8.50	9.20
FF	0.79	0.76	0.76

* Manufacturer's efficiency

Table 2 shows the mass of deposited particulate matter (dirt) accumulated on each module, determined from the samples collected, as well as the mass per unit area for each module, taking into account that their effective area is 1.5120 m^2.

Table 2. Mass of accumulated dust in each PV module evaluated.

Module	Determined Mass (g)	Mass per unit area (g/m^2)
959	0.8385 ± 0.0012	0.5545
979	0.7762 ± 0.0013	0.5133
899	0.5346 ± 0.0016	0.3535
983	0.6379 ± 0.0019	0.4218
918	0.9325 ± 0.0013	0.6166
893	0.5410 ± 0.0015	0.3577
018	0.2801 ± 0.0012	0.1852
834	0.3333 ± 0.0014	0.2204

Based on the results obtained, a graphic of performance loss versus grams of dust deposited on the module per unit area was created (Fig. 6). The efficiency loss is determined by subtracting the actual instantaneous efficiency of the PV module under dirty conditions from that of the clean PV module.

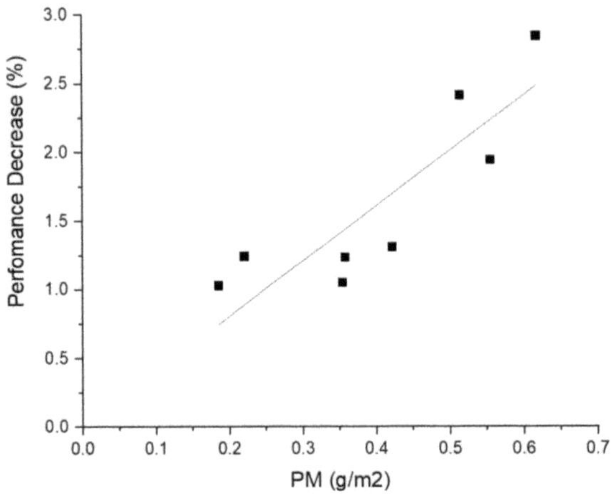

Fig. 6. Decreased performance vs deposited particulate matter on the PV module surface

From the graphic (Fig. 6) of the measuremed values, the relationship between the decrease in performance and the grams of particulate matter deposited on the PV module per unit area was obtained:

$$\text{PerDecr}_{(\%)} = 4.029_{\left(\% / \left(^{g}/_{m^2}\right)\right)} \cdot \text{PM}_{\left(^{g}/_{m^2}\right)} \tag{3}$$

where:

The slope uncertainty is 0.29 (%/(g/m^2)) for a 7.2% uncertainty

$R^2 = 0.9673$

Pearson coefficient = 0.9835

These results show that the greater the mass per unit area of deposited particulate matter on the PV module surface, the greater the performance decrease. An initial approximation based on the experiments performed can be estimated using the equation obtained.

The results show that for an average mass of deposited particulate matter of 0.4029 g/m^2 (~0.4 g/m^2), the performance of the PV modules was reduced by 1.6%. This reduction may seem very small when compared to 100, but the efficiency reported by the manufacturer of the modules used is 19.85%, so a 1.6% reduction means an 8.4% decrease in PV module performance due to the deposited particulate matter.

According to the WMO [14], precipitation or heavy rain is classified as equal to or exceeding an accumulated amount of 100 mm in 24 h, 50 mm in 12 h, or 30 mm in one hour. Several soiling models assume complete cleaning through rain for daily rain sums above a model specific threshold and no cleaning otherwise. However, various studies show that cleaning is often incomplete [15]. During the PV modules' exposure, they were exposed to more than 432.30 mm and 800 mm of rain respectively, with several episodes of heavy rain, for example, 75 mm in 15 min and 50.1 mm in 30 min were measured on April 29 and May 1. These rains could have been considered to clean the panel; however, the decrease in performance is significant due to the accumulation of deposited particulate matter, so planned cleaning is required, using mechanical or chemical means, which should depend on the environment and climate of the place where they are located.

4 Conclusions

- The accumulation of deposited particulate matter on the PV module depends on the environmental conditions where it is located, which include the climate, the natural conditions of the panel, and emissions from human activities.
- Deposited particulate matter on the PV modules reduces their performance and therefore the amount of energy delivered. An average mass of deposited particulate matter of 0.4 g/m^2 decreases PV module performance by 8.4% in the case of half-cell monocrystalline silicon modules.
- In all cases studied, after cleaning, all the PV module's characteristic parameters increased, demonstrating an increase in performance.
- Rain alone does not provide optimal PV module cleaning; planned cleaning is required, which must depend on the environment and climate of the location where the panels are located.

– The greater the mass per unit area of deposited particulate matter on the PV module, the greater the loss in PV module performance, which can be estimated using the equation obtained.

References

1. González, E., Hernández, I.L., López, R., González, C.M.: Dispersión de contaminantes atmosféricos en La Habana de 1994 a 2020. Cub@: Medio Ambiente y Desarrollo, p. 10 (2023)
2. Angulo, J.: Estudio del efecto del polvo y estimación de la portencia nominal en un string fotovoltaico. TECNIA (2020)
3. Rashid, M., Yousif, M., Rashid, Z., et al.: Effect of dust accumulation on the performance of photovoltaic modules for different climate regions. Heliyon **9** e23069, 13 (2023)
4. Bergin, M.H.: Large reductions in solar energy production due to dust andparticullate air pollution. Environ. Sci. Technol. Lett. **4**, 339–344 (2017)
5. Rusănescu, C.O., Rusănescu, M., Istrate, I.A., Constantin, G.A., Begea, M.: The effect of dust deposition on the performance of photovoltaic panels. Energies **16**, 6794. 20 (2023)
6. Amusan, J.A., Muzan, A.I.: The effect of dust on the output performance solar module. Int. J. os Sustain. Energy Environ. Res. 110–122 (2020)
7. Kudusov, M.A., Madvaliev, U., Muqumov, A.R., et al.: The effect of dust deposition on the performance of solar panels in the conditions of the City of Dushanbe. Appl. Sol. Energy **60**, 566–573 (2024)
8. Rodríguez Milián, Y.C., Montero Laurencio, R., Batista Nuñez, Y., Osorio Laurencio, L.: Estudio experimental del efecto del polvo sobre los paneles fotovoltaicos. Ciencia & Futuro, pp. 563–582 (2024)
9. Anbazhagan, G., Usha, S., Santhakumar, J., Surender Reddy, S.: Analysis of dust accumulation effects on the long-term performance of solar PV panels. AIMS Energy **13**(3), 493–516 (2025)
10. Vaillant Roca, L., et al.: Electroluminiscencia de módulos de silicio policristalino. Ingeniería Energética **43**(2), 10 (2022) ISSN:1815–5901 mayo/agosto.
11. IEA PVPS: Review on Infrared and Electroluminiscence imaging for PV fields applications. IEA (2018)
12. IEC TS 62446-3: Photovoltaic (PV) systems–Requirements for testing, documentation and maintenance–Part 3: Photovoltaic modules and plants–Outdoor infrared thermography. Edition 1.0 (2017)
13. IEC 60891.: INTERNATIONAL STANDARD Photovoltaic devices–Procedures for temperature and irradiance corrections to measured I-V characteristics. Edition 2.0 (2009)
14. OMM.: Vocabulario meteorológico internacional. Ginebra: Organización Meteorológica Mundial (1992)
15. Norde Santos, F., et al.: Cleaning of photovoltaic modules through rain: experimental study and modeling approaches. Sol. RRL **8**, 2400551. 10 (2024)

Evaluation of a Cycle of Refrigeration Combining Pinch Analysis with the Exergetic Method

Yanet Sariego Toledo(✉), Eduardo García Noa, Sandra Pérez Ybañez, and Junior Lorenzo Llanes

Universidad Technological de La Habana José Antonio Echeverría (CUJAE), Ave. 114 No 11901, 19390 Marianao, La Habana, Cuba
ysariego@quimica.cujae.edu.cu

Abstract. The present work was developed in a dairy plant and the main objective was to evaluate the refrigeration cycle of the process for ice cream elaboration, considering the exergetic method combined with de Pinch analysis. The exergy losses were calculated applying technical of Process Analysis, it allowed to identify the critical stages for exergeticals relationships and the operation indicators of the refrigeration cycle in the plant. The Aspen Energy Analyzer V7.0 was used to obtain the Big Compound Curves (CCG) and the Big Compound Curves of Exergy (CCGE) like part of the Analysis Pinch to characterize the exergetics behavior of the equipments where heat was exchanged. As fundamental result it was obtained that the refrigeration system works with a operation coefficient of 1.75; a relative yield of 58.35% and a loss of work capacity of 185.67 kW. An alternative is presented that could reduce the losses of work capacity in 14.28% and the compression work in 40.54 kW.

Keywords: Analysis pinch · Aspen Energy Analyzer · exergy · refrigeration systems

1 Introduction

In the dairy plant were the study was carried on, the process of ice cream elaboration is one of the most energy consumer, mainly for the refrigeration processes, for that, to not incur in unnecessary energy expenses is one of the guidelines to fallow.

There is the necessity in the industry to arises a new perspective in the energy uses, focus to the evaluation of the technologies and the introduction of saving technical in the industrial facilities that allow to make more competitive the prices and elaborated products [1, 2].

The traditional method of improve the energetics efficiency as source of saving and less polluting deal with quantify the energy by means of the energy balances, consumptions indexes, inventories and reduction of the losses are used [3]; but also could be use technical of superior order as the exergetics analyses and Pinch analyses.

A. Santos Fuentefria et al. (Eds.): JIFREEEMA 2025, LNNS 1933, pp. 103–113, 2026.
https://doi.org/10.1007/978-3-032-23028-7_10

Due to the thermodynamic evaluation of refrigerating facilities occupies today an important place for the professionals, the present paper has as objective to evaluate exergeticaly the refrigeration cycle of the process ice creams elaboration by means of the exergetic analysis and the Pinch analysis [4, 5].

2 Materials and Methods

The procedure for the evaluation of the refrigeration system for the elaboration of ice creams, was developed by means of the following steps:

2.1 Description of the Refrigeration System for the Elaboration of Ice Creams

The system is a cycle of refrigeration of double stage of vapor compression that uses the ammonia as regrigerant, which is as saturated vapor in the tank of low pressure (TKB), just as it is shown in the Fig. 1, at 64.4 kPa and -42 °C for a compressor screw (CPB). Later, it is compressed and send as overheated vapor to the intermediate tank (TKI).

Having bubbled through the ammonia liquid present in the tank, it cools down and it became in a saturated vapor. After that, it is discharged in the intermediate pressure channel where it is again extracted as saturated vapor (pressure of 291.3 kPa and temperature -10 °C) for a reciprocal compressor (CPA) in the high pressure section reaching the overheated vapor state, which is send to the condenser (tubes and shell exchange with 1 357 kPa and temperature 119 °C). In this equipment it exchange heat with treated water that it is recycled through the cooling towers and condense.

The ammonia liquid obtained in the condenser goes toward the receiver tank of liquid (TKL) from which leave two currents (VEX-1, VEX-2), but both of them, have a fall of pressure before they arrive to its destination that causes, on one hand, its partial evaporation and for the other one, the temperature of evaporation of the remaining liquid is reached. One of these currents is send to the receiver intermediate tank (-10 ºC and 291.3 kPa) to provoque the cooling of the overheated vapor discharged from the low pressure compressor. While the other one is send to the ice bank (BCH), where exchanges heat with the current of freeze water that coming from the ice cream plant. In the ice bank the ammonia is evaporated to maintaining the temperature at 1 ºC, that is to say, in this stage it happens the refrigeration of the water properly, which is used in several equipments in the process of ice cream elaboration.

After the ammonia evaporation in the ice bank serpentine it is discharged in the intermediate receiver tank. The ammonia liquid in this separator and cooler tank goes to the low pressure receiver tank (TRB) (64.4 kPa and temperature -42 °C) due to the signal of a solenoid valve, then it is pumped to the refrigerator system (NEV) and freezers (COG). There is an expansion during this movement through the TKI to the TKB due to the difference of pressures. In the COG the ammonia exchanges heat with the mixture to freeze and in the refrigeration chamber the exchange is with the air impelled by the diffusers. The evaporated ammonia is again sent to the low pressure receiver tank (64.4 kPa and temperature -42 °C) closing the refrigeration cycle.

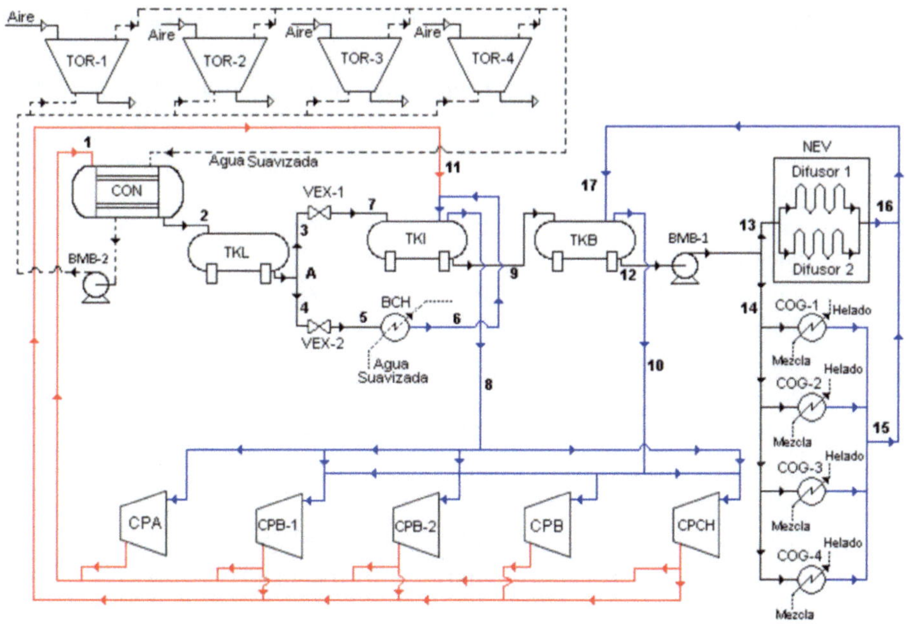

Fig. 1. Diagram of the plant of refrigeration

2.2 Thermodynamic Representation of the Cycle

In the Fig. 4b is represented the cycle in a pressure - entalpy diagram. The process of heat remove was studied during fifteen days, taking messures in the machinary room and in the elaboration of ice cream process. There were operational messures to know the temperature and pressure levels of each working point. Those messures appear in the Table 1. Then were found the entalpy and entropy of the fluids in each working points. The obtained values are also shown in the Table 1.

2.3 System Evaluation by Exergetic Method

The methods more applied in the different industrial sectors, including the nutritious industry, to determine the thermodynamic behavior of a system and to propose improvements is based on thermodynamic first law.

The main limitation of the energy methods of analysis resides in that it is focus in the materials and energy balances to find and reduction the energy losses, for the determination of consumption indexes and operation indicators, as well as in energy audits. However they don't establish differences for the quality of the energy that is distributed or how is the quality of the energy transformations.

Table 1. Thermodynamic properties of the refrigerant in the refrigeration cycle

Current	State	Pressure (kPa)	Temperature (ºC)	Entalpy (kJ/kg)	Entropy (kJ/kg K)
1	overheated vapor	1357	119	1721.29	5.88
2	saturated liquid	1357	35	362.36	1.55
3	saturated liquid	1357	35	362.36	1.55
4	saturated liquid	1357	35	362.36	1.55
5	wet mix	291.3	-10	362.36	1.62
6	saturated vapor	291.3	-10	1449.13	5.75
7	wet mix	291.3	-10	362.36	1.62
8	saturated vapor	291.3	-10	1449.13	5.75
9	saturated liquid	291.3	-10	154.31	0.83
10	saturated vapor	64.4	-42	1404.23	6.28
11	overheated vapor	291.3	63	1621.35	6.33
12	wet mix	64.4	-42	154.31	0.83
13	wet mix	64.4	-42	154.31	0.83
14	wet mix	64.4	-42	154.31	0.83
15	saturated vapor	64.4	-42	1404.23	6.28
16	saturated vapor	64.4	-42	1404.23	6.28
17	saturated vapor	64.4	-42	1404.23	6.28
Reference state		101.3	30	1557.58	6.636

The quality of the energy is related with the capacity to carry out useful work. The maximum quantity of useful work is obtained when the system carries out a reversible process where only exchanges heat with the surroundings. The maximum useful work is also known as work capacity or disponibility [6–9].

The exergy is defined as the maximum quantity of attainable work when an energy flow is taken from its initial state to an equilibrium thermodynamic state with the substances in their natural environment (inert state), by means of reversible processes, interacting only with his natural environment [10]. If the nuclear, magnetic and electric effects are excluded, it is possible to differentiate four fundamental types of exergy: physical, kinetic, potential and chemical exergy [7].

In this work it was considered the physical exergy only because:

- The kinetic exergy and potential are negligible
- The chemical exergy is zero (chemical reaction doesn't take place)

The physical exergy (*e*) was calculated from the following general expression (1):

$$e = (H - H_0) - T_o(S - S_0) \tag{1}$$

where *e* represents the physical exergy (kJ/kg), H and H_0 entalpy of the initial state and reference respectively (kJ/kg), S and S_0 are the entropy of the initial states and reference respectively (kJ/K) and T_0 temperature of the reference state (K).

Often during the exergy analysis of a process it is more necessary to know the difference between physical exergy beetwen two states of the same sustance than with the reference state [7], in such a case the expression for the calculation is the Eq. (2):

$$e = (H_2 - H_1) - T_o(S_2 - S_1) \tag{2}$$

The capacity of work of the energy or the maximum useful work of the heat due to the interaction with the environment is the exergy of the flow of the heat. In general this exergy of the heat can be evaluated by Eq. (3):

$$e_q = \left(1 - \frac{T_0}{T}\right) \cdot Q \tag{3}$$

where e_q represents the exergy of heat (kW), T is the temperature (K) and Q absorbed heat of the hot focus (kW).

One of the difficulties for the application of the exergetic method in the refrigeration systems is the definition of the reference state temperature because those systems work indistinctly to superior and inferior temperatures of the environment, the problem of the negative exergies arises in some points in direct conflict with the thermodynamic principles. Some authors solve this problem adding the biggest negative exergy obtained to each flow of heat of the system, changing the scale of the reference system and transforming all the average negative exergetics into positive terms [8–11], this procedure will be used in the present investigation.

2.4 Losses of Work Capacity

The losses of work capacity in the refrigeration systems are due to the existent deviations in the real cycles regarding the ideal cycle [12]. The analysis method derived from the second law is known as the exergetic method and it consists on to evaluate the effectiveness of the process by means of the loss of the work capacity, which can be determined by means of the following Eq. (4):

$$\Delta L = \left[\left(e_{inlet} + e_q\right) - e_{oulet}\right] - W \tag{4}$$

where L is the loss of work capacity (kW), e_{inlet} and e_{oulet} are the exergy of the entrance and exit flow respectively (kW), and W compression work (kW).

For energy evaluation was used the refrigerant thermodynamic properties obtained in the diagram P-H represented in the thermodynamic cycle that is shown in the Table 2.

It was carried out the exergetic balance for each component of the system and the reference state was defined by the ambient temperature and the atmospheric pressure whose values are respectively 301 K and 101.3 kPa.

Table 2. Currents that exchange heat in the refrigeration cycle

Currents	Temperature (K)		Caloric capacity (kW/K)	ΔH (kW)
	Inlet	Outlet		
Ice cream to COG	277	267	1.84	18.40
Water to BH	280	274.5	6.63	36.49
NH_3 to Condenser	392	308	9.06	761.00
Air to NEV	305	246	1.11	65.62
Utilities				
Cooling tower water	296	308	-	765.4
Refrigerant 1($NH_{3(I)}$)	263	263	-	89.65
Refrigerant 2 ($NH_{3(II)}$)	231	231	-	26.31

On the other hand it was necessary to estimate the entalpy and the entropy of the semi frozeen mixture and frozeen ice cream. The variation of exergies of flow in the evaporators regarding the flow of the inlet components is evaluated then by the following Eq. (5):

$$\Delta e_{evaporador} = \sum x_i \cdot \Delta e_i \tag{5}$$

where e_i is the variation of exergy of flow of each component (kW) and x_i the mass fraction of each component in the mixture.

Once determined these exergy variation, the loss of work capacity is evaluated by means of the Eq. (4) but in this case it is added the variation of exergy due to the flow of the product current.

2.5 Analysis Pinch Procedure for the System Evaluation

One of the tools that has demonstrated their effectiveness to treat problems of optimal design, as well as in the redesign of exchange of heat system is the Analysis Pinch (AP) whose bigger advantage is the simplicity of the concepts in which it is based. The tools better well-known of the AP are the compound curves (DC), where the thermal characteristics of the different currents of the process are represented, showing the energy requirements (cooling/heating) of those. Another tool of the AP is the Great Compound Curve (GCC). The GCC presents a profile T-H of the external requirements of heating and cooling after the recovery of heat [4, 5], offering a clear image of the interface between the process and the system. The results obtained were given in terms of thermal loads from the DC and the GCC.

In the DC the area among the hot and cold curves represents the exergy loss; however in a diagram T-H this area is not similar to the exergy losses. In systems like the refrigeration cycles, where are present thermal machines the compound curves should be modified. Linnhoff and Dhole [13], introduced a method based on the exergy to estimate the required work using the data of the currents. In the procedure it is possible to obtain the curves of exergy (CCE) and the great curve of exergy (GCCE) transforming the axis of the temperatures of the DC and the GCC in the factor of Carnot ($\eta = 1\text{-}T_0/T$).The area among the curves represent the losses exergetics.

To obtain the GCCs of the cycle, as well as the GCCs with the utilities, it was employee the Aspen Energy Analyzer V7.0 taking a minimum difference of temperature of 5ºC. The data for the construction of the curves are shown in the Table 2 [14].

3 Results and Discussion

In the processes of heat exchange the exergetic losses depend on the temperature difference and the exchanged thermal load. In the Fig. 2 (a) the GCC of the process is shown with the utilities of the refrigeration system, where the biggest differences of temperature between the process currents and the utilities are in the condenser, causing the high values of losses of work capacity.

An alternative to improve the behavior of the cycle is to reduce the temperature difference among the currents in the exchange [3, 15]. This alternative will discusses later for the case of the interchanger. In the Fig. 2b the GCCE is shown, where is observed that it is possible to increase the exergetical efficiency in the refrigerators chamber, increasing the temperature of the exchange medium. Nevertheless, although theoretically it is possible, would not be satisfy the refrigeration requirements that the process demands.

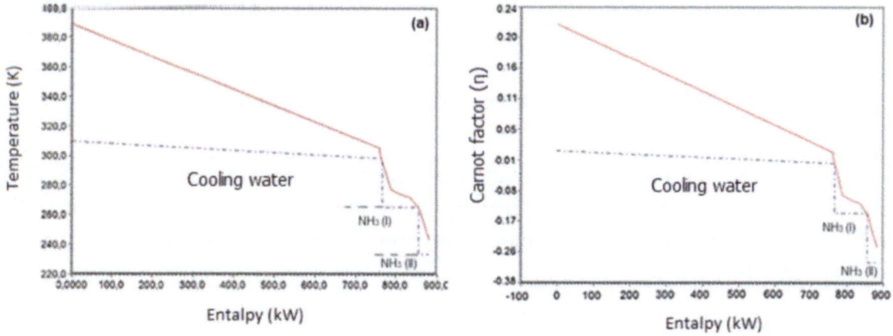

Fig. 2. GCC of the plant of refrigeration (a) y GCCE of the plant of refrigeration (b)

The results of the exergetic method of determination of the losses of work capacity for each element of the cycle are shown in the Fig. 3. It could observe that the equipments that present the biggest losses of work capacity are the condenser and the receiver tank fallowed in magnitude by the three evaporators (refrigerator chamber, freezer and bank of ice).

The real cycles of refrigeration for mechanical compression of vapor differ from the ideal cycles in the irreversibilities that are generated due to the recirculating refrigerant friction and for the temperature differences between the refrigerant and the fluids that exchanges heat and with the reference state too. The irreversibilities cannot be avoided, but they should be diminished to a minimum value, because those cause to suply additional power, to achieve the wanted refrigeration.

3.1 Alternative to Increase the Exergetic Efficiency

As a decrease in the temperature difference allows to increase the exergetic heat exchange efficiency, it was studied to diminish the condensation temperature in 5 °C. That produced a decrease in the area between the GCCE and the level of the utility, just as it is shown in the Fig. 4a, reducing the exergetic losses.

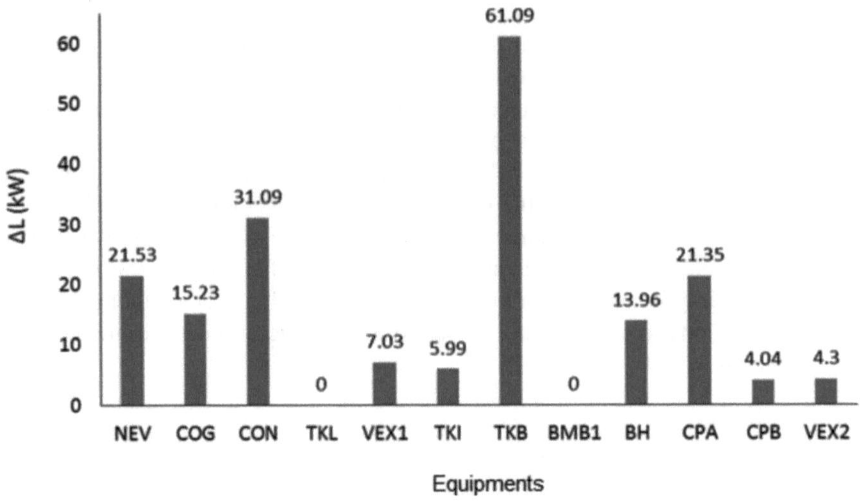

Fig. 3. Losses of work capacity for each equipments

In the Fig. 4b, the refrigeration cycle has been represented in the diagram P-H for the new work conditions and the thermodynamic properties of the refrigerant in this conditions. Then, the temperature and the pressure of the working points 1, 2, 3, 4, 5 and 7 are modified.

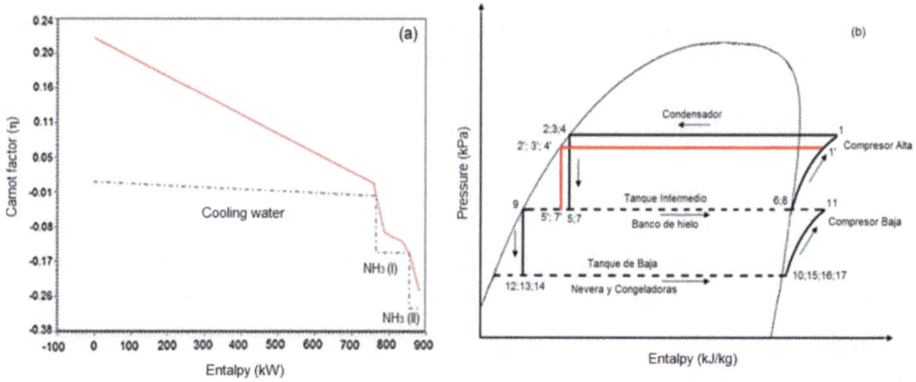

Fig. 4. The GCCE for the alternative proposal (a), and diagram P-H of the cycle of refrigeration of the base case (black line) and the alternative (red line) (b).

The alternative reduce the losses of work capacity in 26.52 kW, that represents a reduction of 14.28% regarding the base case, improving the exergetic efficiency of the refrigeration cycle. In addition, it causes an integral increment in the energy efficiency of the cycle, because improvement the refrigeration efficiency and improves the quality of the energy transferred in this processes too.

The alternative is favorable from the thermodynamic point of view; similar results were obteined in energetic analysis [2, 15–17]. The results are shown in the Fig. 5.

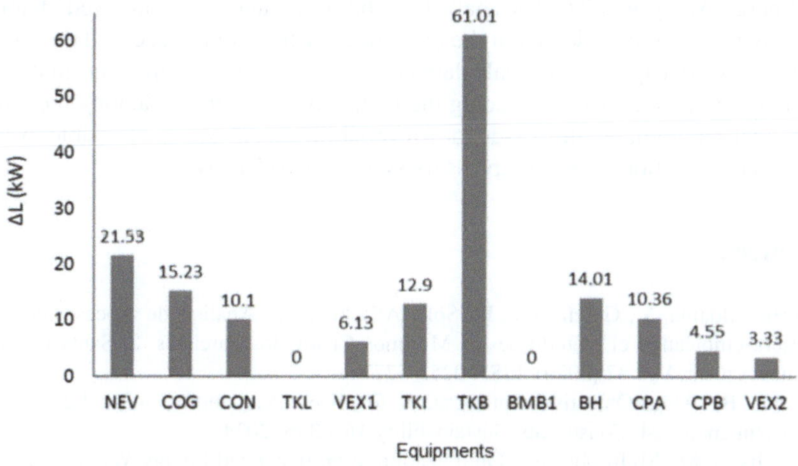

Fig. 5. Losses of the capacity of work of the alternative

In the Table 3 is shown a comparison among some operation indicators of the refrigeration cycle of the factory. It is important to notice that the compression work on the low pressure area increases, while there are a decrease of the work compression in high pressure area, resulting in an increase of the coefficient of operation of the real cycle.

That result show that with this alternative increase the energy efficiency of the cycle, reducing the consumption of power in 40.54 kW [18].

Table 3. Comparison of indicators of the refrigeration cycle of the plant

Indicator	High pressure zone		Low pressure zone		System	
	Base Case	Alternative	Base Case	Alternative	Base Case	Alternative
W (kW)	153,07	105,49	56,08	63,12	209,15	168,61
COP real	1,52	2,20	2,39	2,12	1,75	2,17
COP Carnot	5,84	5,84	3	3	3	3
Relative performance (%)	26	70,68	79	37,71	58,35	72,38

4 Conclusions

The results obtained with the application of the exergetic study of the refrigeration cycle of the factory demostrated that it, there is possibilities to improve the process conditions to increase the energy eficience, based on the combination of the techniques of the analysis Pinch and the exergetic method. The GCC of the cycle was obtained using the Aspen Energy Analyzer V7.0. The analysis of the compound curves allowed identifying that the biggest exergetics losses in the exchanger of heat took place in the condenser. The losses of work capacity were calculated for each one of the components of the cycle. It was analyzed an alternative reducing the temperature difference among the currents in the condenser, reducing the losses for irreversibilities. Also it was possible to reduce the power consumption in the compressors system in 40.54 kW.

References

1. Barrera Aldama, Y., García Noa, E., Solis Aliaskina, K.: Análisis de procesos en la cuba quesera empleando el método de los Mínimos Cuadrados Parciales. 2, Santa Clara: s.n., Centro Azúcar, Vol. 47 (2020). ISSN 0253-5777
2. Hassna, B.: Multi-Objective Optimization for Food Availability under Economic and Environmental Risk Constraints. Sustainability **16**, 4336 (2024)
3. Arif-Misrol, M.: Multi-objective Optimization of an Integrated Energy-Water-Waste Nexus for Eco-Industrial Park. Malaysia: s.n., Chemical Engineering, Vol. 89 (2021). ISBN 978-88-95608-87-7; ISSN 2283-9216
4. Smith, R.: Chemical Process Design and Integration. Segunda ed. Wiley (2016)
5. Smith, R.: State of art in process integration. Applied Thermal Engineering **20**, 1337–1345 (2000)
6. KIrillin, V.A., Sichev, V.V.: Termodinámica térmica. Segunda Edición ed. Editorial MIR, Moscú (1986)
7. Kotas, T.J.: The Exergy Method of Thermal Plant Analysis. Butterworths, London (1995)

8. Gordon, J.: Cool Thermodynamic's. Cambridge International Science Publishing, UK (2000)
9. Khan, J.: Thermodynamic optimization of finite time vapor compression refrigeration system. Exergy Conversion and Managgement **42**(3), 145–148 (2001)
10. Szargut, J., Morris, D.R., Steward, F.R.: Exergy Analysis of Thermal, Chemical, and Metallurgical Processes. Hemispliere Publishing Corporation, New York (1988)
11. Yumrutas, R.: Exergy analysis of vapor compression refrigeration system. Exergy an Int. J. **2**(2), 266–272 (2002)
12. Armas, J.: Análisis exergético de un sistema de climatización centralizado por agua helada. Revista Energética **36**(3), 35–40 (2006)
13. Linnhoff, B., Dhole, V.R.: Shaft work targets for low temperature process design. Chem. Eng. Sci. **4**(8), 2081–2091 (1992)
14. Ramírez-Martín, C.: Simulator to improve the productivity of yogurt processing based on a model of finite automaton. Int. J. Eng. Technol. **11**(2), 344–351 (2019)
15. Triana Pujol, A.: Estudio del comportamiento de indicadores de desempeño energético de un hotel de Varadero. 1, La Habana, Cuba: s.n., Vol. 44. Ingeniería Industrial. ISSN 1815-5936 (2023)
16. Valera Sterling, E.C., García Noa, E.: Empleo de fuentes renovables de energía y métodos de integración de procesos en la industria láctea. La Habana, Cuba: s.n. Ecosolar, **83**, 29–35 (2023). ISSN-1028-6004
17. Bergamini, R.: Analysis of energy integration opportunities in the retrofit of a milk powder production plant using the Bridge framework. J. Clean. Prod. **328**, 129402 (2021)
18. Kunal, D., Choudhary, R., Thompson-Witrick, K.A.: Effects of new technology on the current manufacturing process of yogurt-to increase the overall marketability of yogurt. LWT **108**, 69–80 (2019)

Characterization of the Renewable Energy Generation Potential of Cayo Carenas

Boris G. Vega Lara[1(✉)], José P. Monteagudo Yanes[1], and Royd Reyes Calvo[2]

[1] Center for Energy and Environmental Studies, Faculty of Engineering, University of Cienfuegos, Carretera a Rodas Km. 4, Cuatro Caminos, Cienfuegos, Cienfuegos, Cuba
{bvega,jpmyanes}@ucf.edu.cu

[2] Department of Mechanical Engineering, Faculty of Engineering, University of Cienfuegos, Carretera a Rodas Km. 4, Cuatro Caminos, Cienfuegos, Cienfuegos, Cuba
royrc@ucf.edu.cu

Abstract. This article characterizes the renewable energy potential of Cayo Carenas, an islet at the entrance of Cienfuegos Bay, Cuba. It begins with a concise overview of the case-study community's geographical and socio-economic context. To assess solar and wind potential, the concepts of Peak Sun Hours (PSH) and Wind Power Density (WPD) are introduced, alongside the most utilized probability distributions in wind energy studies. Solar irradiance and wind speed data, collected at 10-minute intervals throughout one year by an automatic weather station installed near the islet by Cienfuegos Provincial Meteorological Center, are analyzed. Daily PSH and WPD averages are computed, followed by derivation of monthly means for the study period. Frequency histograms for PSH and wind speed are generated. Results confirm exceptional annual averages: solar potential exceeding 5 PSH and wind potential higher than 400 W per square meter WPD. Short-term solar deficits are identified during cloudy/rainy periods, while below-average wind potential occurs from June-September. Weibull and Rayleigh distribution parameters, estimated via Maximum Likelihood Method, effectively model the stochastic behavior of wind speeds.

Keywords: solar and wind energy potential · isolated community

1 Introduction

A core pillar of microgrid design and implementation involves integrating high penetrations of renewable energy sources (RES) into power supply schemes. Microgrids' inherent capabilities, energy storage and advanced management systems, enable effective handling of intermittent, non-dispatchable RES like solar

J. P. M. Yanes and R. R. Calvo—These authors contributed equally to this work.

A. Santos Fuentefria et al. (Eds.): JIFREEEMA 2025, LNNS 1933, pp. 114–125, 2026.
https://doi.org/10.1007/978-3-032-23028-7_11

and wind. This transforms them into providers of stable, continuous, and reliable electricity [1].

Effective design of high-RES-penetration power systems necessitates a holistic assessment of available energy resources within the target territory, accounting for site-specific environmental characteristics [2]. This requires comprehensive energy potential characterization, quantifying convertible energy from primary sources into heat or electricity. Consequently, meticulous resource prospecting becomes imperative prior to selecting optimal RES technology combinations, with cost-reliability trade-offs informing territorial deployment strategies [1].

Photovoltaic (PV) panels and wind turbines (WT) constitute the most prevalent renewable energy sources in standalone hybrid systems, primarily in two configurations: PV-WT-Battery and PV-Generator-Battery. Lead-acid batteries dominate due to their relatively low cost, though alternative electrical storage technologies exist (lithium-ion batteries, supercapacitors) [1].

The foundational step in microgrid design involves determining the target site's suitability for PV and WT installations through comprehensive renewable energy potential assessment (solar/wind). The most prevalent methodologies employ meteorological data and statistical analyses, necessitating 1–2 years of measured parameters: solar irradiance, wind speed/direction. While temperature and relative humidity enhance modeling accuracy, they are non-essential for fundamental potential quantification [1].

Research on methods and tools for assessing renewable energy source availability has been extensively documented [2–4]. Electricity generation potential from solar energy has been investigated in references [5–11]. Notable Latin American studies include: Martins et al. present a solar resource assessment during the Solar and Wind Energy Resource Assessment (SWERA) project, utilizing the BRAZIL-SR radiation transfer model with satellite and climatic data for Brazil [6]. Ordoñez et al. evaluate solar and wind resources for Bucaramanga, Colombia, based on meteorological records from four strategic monitoring points during 2010–2012 [8].

Within the European continent, the following contributions are prominent: Šúri et al. investigate the theoretical electricity generation potential of photovoltaic systems across the European Union [5]; Makrides et al. analyze the performance potential of diverse PV configurations and compare their energy yield in two high-insolation European countries [7]; Perpiña Castillo et al. assess the solar energy potential of European regions using Geographic Information Systems (GIS), developing high-spatial-resolution photovoltaic suitability maps for the European Union [10].

Ramgolam and Soyjaudah assess the solar energy potential of Mauritius, utilizing in-situ measured solar radiation data supplemented by specialized software outputs and international databases [9]. He and Kammen employ a decade of hourly solar radiation data from 200 locations to develop solar availability profiles across China [11].

In Cuba, notable contributions include [12–14], among others. Stolik analyzes multiple aspects of photovoltaics environmental impact, manufacturing, markets, costs, dominance of crystalline silicon cells, economic assessments, and short-to-long term PV energy forecasts for Cuba [12]. Rodríguez Gámez et al. evaluate renewable energy potential (solar, wind, hydro, biomass) across Havana and Santiago de Cuba provinces using GIS, targeting sustainable development goals in studied communities [13]. González Lorente et al. examine the impact of different regulatory frameworks on RES integration in Cuba's and Spain's power generation sectors [14].

For wind energy potential assessment, notable references include [8,15–22]. Ilinca et al. present the development of a comprehensive wind atlas for Québec Province, Canada [15]. Castellanos and Ramesar estimate Trinidad and Tobago's wind resources using autoregressive models and probability density functions [16]. Migoya et al. evaluate wind potential through wind atlas development for the Madrid Region, Spain [17]. Carta et al. provide a review of probability distributions for wind energy analysis, using the Canary Islands as a case study [18].

Ucar and Balo evaluate wind energy potential for turbine installations in Turkish coastal regions [19]. Ouammi et al. characterize monthly and seasonal wind energy patterns across four Italian locations [20]. Ahmed analyzes wind potential at Ras Benas, Egypt, using measured meteorological data and the Weibull probability distribution [21]. De Araujo Lima and Bezerra Filho assess wind resources and simulate a wind farm in Triunfo, Pernambuco, Brazil [22].

In Cuba, Torres Durán and Moreno Figueredo assess wind potential in Cojímar, Havana [23]. Building on wind resource analysis at Las Terrazas complex, Moreno Figueredo proposes a turbine sizing and selection model for water pumping in piedmont zones [24]. Sánchez Torres and Rodríguez Ramos evaluate wind energy potential at 10 m/30 m heights, statistically analyze wind speed/direction and power density, and conduct techno-economic feasibility studies for small turbines in Havana [25].

This study reports the characterization of renewable energy generation potential (solar and wind) for Cayo Carenas islet, located at the entrance of Cienfuegos Bay, Cuba. The contributions of this article are: (i) determination of average daily peak sun hours (PSH) for all months of a year, along with the annual mean; (ii) calculation of average daily wind power density (WPD) for all months of a year, along with the annual mean; and (iii) estimation of Weibull and Rayleigh probability distribution parameters describing wind speed behavior at the case-study location.

This article is structured as follows. Section 2 presents a concise overview of Cayo Carenas' geographical and socio-economic context, along with the meteorological variables employed. Section 3 introduces the concept of Peak Sun Hours (PSH) for solar energy potential assessment. Section 4 defines Wind Power Density (WPD) and discusses the most utilized probability distributions in wind power studies. Section 5 details the renewable energy generation potential characterization results for the case-study community. Section 6 concludes with the study's key contributions.

Fig. 1. Spatial configuration of Cayo Carenas within Cienfuegos Bay. Source: [26].

2 Case-Study Community: Cayo Carenas

The islet of Cayo Carenas is situated at the entrance of Cienfuegos Bay on Cuba's southern coast. Its geographic coordinates are 22°5'13" N, 80°27'36" W, with an approximate area of 20 hectares. Figure 1 shows the spatial configuration of the islet within the bay.

The islet contains a significant archaeological site of pre-Columbian agro-ceramic inhabitants. Historical records document pirate settlement activities during the early 16th century. Additionally, the location served as a filming site for an acclaimed Cuban motion picture [27].

With a permanent population of 10–15 residents, the islet contains 14 occupied dwellings. The primary economic activity involves fishing and commercialization of marine products. Electricity service, provided free of charge, is supplied by a 12 kW diesel generator operating 10–12 hours daily, consuming 600–620 liters of diesel monthly [28].

The islet lacks both water supply and sewerage systems. Potable water is harvested in cisterns and external storage tanks, primarily sourced from rainfall. Food preparation relies on LPG (liquefied petroleum gas) stoves or charcoal. Despite having no signal repeaters or telephone base stations, the location receives radio broadcast, television transmission, and mobile network coverage with internet data services. Overall, living conditions on Cayo Carenas remain challenging due to substandard housing infrastructure, limited access to electricity/water services, and unreliable marine transportation to adjacent settlements. These constraints drive continuous outmigration of residents [28].

2.1 Weather Data

To establish a reliable forecast of renewable energy potential (solar or wind) for a case-study location, at least one year of site-specific meteorological data is required [2]. For Cayo Carenas, data was provided by the Provincial Meteorological Center (CMP) of Cienfuegos. Measurements were collected at 10-minute intervals throughout 2011 using an automatic weather station installed near the islet. Technical failures at the weather station prevented data acquisition in subsequent years.

The recorded meteorological variables include: ambient temperature, global solar irradiance, wind speed/direction, humidity, dew point, atmospheric pressure, precipitation, among others. Figure 2 displays the 2011 daily profiles for solar irradiance and wind speed. Notably, minimal data gaps occurred due to instrumentation malfunctions or severe weather events, collectively representing fewer than six days of missing data.

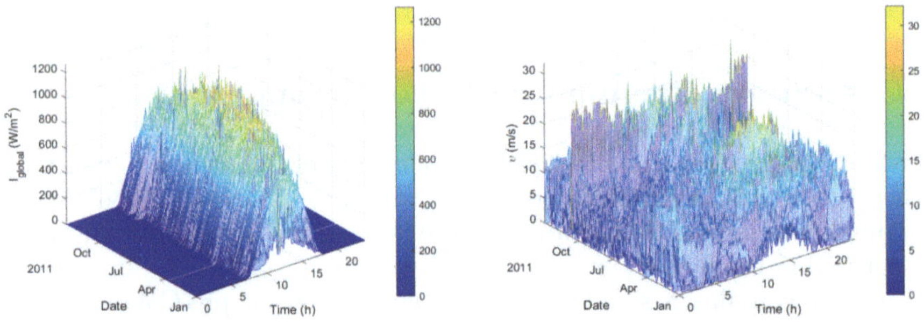

(a) Solar radiation. (b) Wind speed.

Fig. 2. Weather data from the Provincial Meteorological Center of Cienfuegos.

3 Solar Energy Potential

3.1 Peak Sun Hours

Daily solar energy potential is quantified through Peak Sun Hours (PSH) [1]. PSH represents the equivalent daily hours of constant 1000 W/m^2 solar irradiance. Since PV system rated power is defined under Standard Test Conditions (STC), 25 °C ambient temperature and 1000 W/m^2 irradiance, PSH determination provides an efficient and practical method for calculating energy yield from PV modules [29].

The mathematical expression for calculating PSH is given by Eq. (1):

$$PSH = \frac{\int_{t_{sunrise}}^{t_{sunset}} I_{global}\left(t\right) dt}{I_{global,STC}} \tag{1}$$

where $I_{global}(t)$ denotes time-dependent global solar irradiance, $I_{global,STC} = 1000\ \mathrm{W/m^2}$, and $t_{sunrise}$, t_{sunset} represent sunrise and sunset times respectively. Locations with $PSH > 3$ h are considered suitable for electrical power generation using PV systems [30].

4 Wind Energy Potential

4.1 Wind Power Density

One of the most widely adopted approaches for estimating wind energy potential involves determining the site-specific Wind Power Density (WPD) [1,31]. Theoretically, wind power P, where air density ρ, flows through area A, follows a cubic function of wind speed v, as expressed in Eq. (2):

$$P = \frac{1}{2}\rho A v^3 \tag{2}$$

WPD is defined as wind power per unit area, as formalized in Eq. (3):

$$WPD = \frac{1}{2}\rho v^3 \tag{3}$$

Given known wind speed values, the average wind power density, WPD_{avg}, can be calculated for specific time periods (typically months or years) using Eq. (4), where N represents the number of observations:

$$WPD_{avg} = \frac{P_{avg}}{A} = \frac{\sum_{i=1}^{N} 0.5\rho_i v_i^3}{N} \tag{4}$$

Using average ambient temperature and atmospheric pressure values, the mean air density, ρ_{avg} can be determined. Subsequently, WPD_{avg} is calculable via Eq. (5) using wind speed data [31].

$$WPD_{avg} = \frac{P_{avg}}{A} = \frac{0.5\rho_{avg}}{N}\sum_{i=1}^{N} v_i^3 \tag{5}$$

The U.S. Department of Energy (DOE) Wind Energy Program has established a 7-tier classification scale for wind resource assessment using WPD. Sites rated Class 4 ($WPD_{avg} \leq 250\ \mathrm{W/m^2}$, $v_{avg} \leq 6.0$ m/s at 10 m height) or higher are suitable for wind turbine installation. Class 3 locations ($WPD_{avg} \leq 200\ \mathrm{W/m^2}$, $v_{avg} \leq 5.6$ m/s at 10 m) require turbines on taller hub heights, while Class 2 sites ($WPD_{avg} \leq 150\ \mathrm{W/m^2}$, $v_{avg} \leq 5.1$ m/s at 10 m) or lower are not viable for wind energy projects [1,15,31].

4.2 Wind Speed Probability Distributions

When comprehensive annual wind speed measurements are available for the case-study location, statistical tools may be employed to determine the wind's probability density function. Probability distribution modeling provides a streamlined method for wind energy potential assessment, as it yields concise and practical characterization of wind speed frequency distribution [2,18].

The most widely used probability density function for statistical analysis of wind speed data is the Weibull distribution. Its advantages include: high flexibility, dependence on only two parameters, simple parameter estimation across methodologies, strong empirical fit, among others [18].

The Weibull distribution, defined by Eq. (6), is characterized by two parameters. The shape parameter, k, indicates probability distribution width; higher k values describe narrower distributions, reflecting wind speeds concentrated within a limited range. The scale parameter, c, represents wind resource magnitude, quantifying available wind energy [1].

$$f(v) = \frac{k}{c}\left(\frac{v}{c}\right)^{k-1} e^{-\left(\frac{v}{c}\right)^k} \tag{6}$$

A special case of the Weibull probability density function is the Rayleigh distribution, defined by Eq. (7). Here, the single-parameter Rayleigh distribution (scale parameter b) is equivalent to a Weibull distribution with shape parameter $k = 2$ and scale parameter $c = \sqrt{2}b$. Due to its simplicity, the Rayleigh distribution remains widely adopted in wind power studies [18].

$$f(v) = \frac{v}{b^2} e^{\left(\frac{-v^2}{2b^2}\right)} \tag{7}$$

Once the most commonly used probability distributions for wind studies are identified, the next step involves estimating the unknown parameters of these density functions. The most prevalent methods for parameter estimation include: the Method of Moments (MM), the Maximum Likelihood Method (MLM), and the Least Squares Method (LSM). Among these, the MLM has been empirically validated as the most efficient approach for estimating Weibull distribution parameters [18,32].

5 Energy Potential Assessment Results

Initially, PSH were calculated for all days with complete solar radiation measurements (359 total). Figure 3a displays the frequency histogram of PSH for the study period. The histogram reveals few days with values below $3h$, attributable to prevailing cloudy/rainy weather conditions. However, the vast majority of days exhibit high solar energy potential ($> 4h$), characteristic of the islet's tropical latitude. This is further corroborated in Fig. 3b, illustrating consistently high daily average potential across all months, only January and December show values slightly below $4h$.

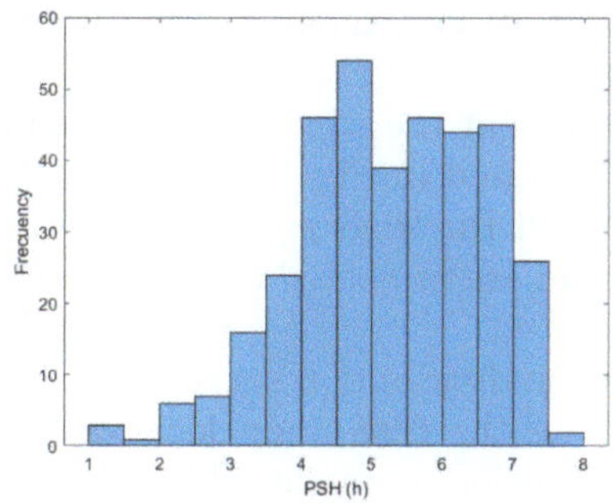

(a) Peak sun hours histogram

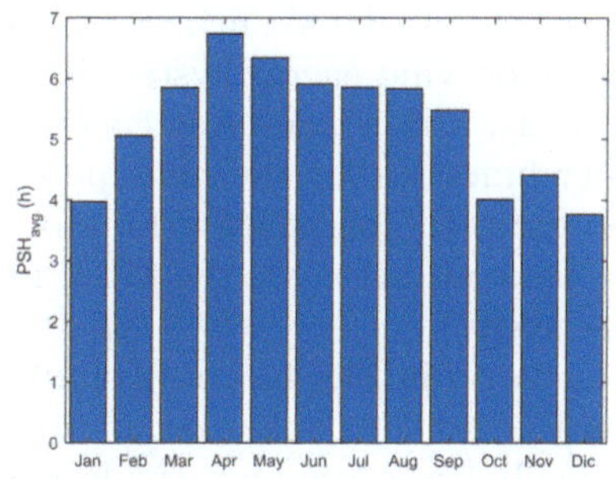

(b) Daily average solar potential by month.

Fig. 3. Solar energy potential statistics: $PSH_{avg} = 5.26$ h; $PSH_{std} = 1.28$ h.

Subsequently, it was determined that the case-study community at Cayo Carenas has a daily average peak sun hours value of $PSH_{avg} = 5.26$ h with a standard deviation of $PSH_{std} = 1.28$ h for the evaluated year. This demonstrates exceptional solar energy potential, highly suitable for photovoltaic system deployment.

For wind energy potential assessment, the daily average WPD was calculated monthly for the study period, as shown in Fig. 4b. The graph reveals that March and April exhibit significantly higher WPD than other months due to persistent and occasionally strong winds associated with the Lenten wind season [33]. Conversely, months from June through September demonstrate substantially lower performance compared to the rest of the year.

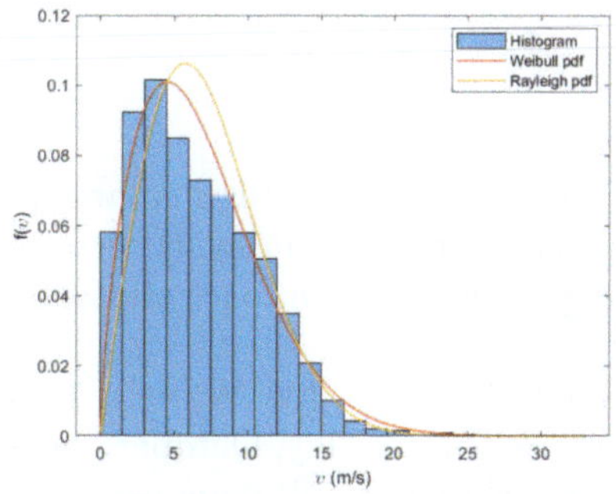

(a) Probability distributions of wind speed.

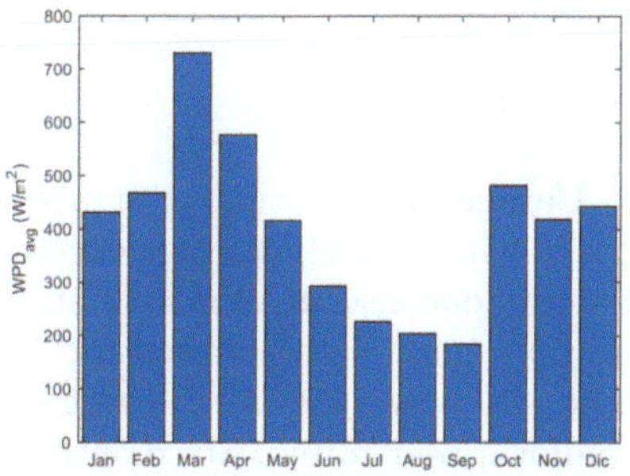

(b) Daily average wind potential by month.

Fig. 4. Wind energy potential statistics: $v_{avg} = 6.3$ m/s; $WPD_{avg} = 406$ W/m^2.

The site is characterized by an average wind speed of $v_{avg} = 6.3$ m/s and an average wind power density of $WPD_{avg} = 406$ W/m^2 for the evaluated year. This places Cayo Carenas within the DOE-proposed wind class scale between Class 5 ($WPD_{avg} \leq 300$ W/m^2, $v_{avg} \leq 6.4$ m/s at 10 m height) and Class

6 ($WPD_{avg} \leq 400$ W/m^2, $v_{avg} \leq 7.0$ m/s at 10 m height), making it highly suitable for wind energy system installation.

Finally, a histogram was generated to represent the wind speed probability density functions for the same period, as illustrated in Fig. 4a. Using the wind speed data, the parameters for the Weibull and Rayleigh probability distributions were estimated via the MLM method, and are presented in Tables 1 and 2, respectively.

As shown in Fig. 4a, the Weibull probability distribution successfully reproduces the random behavior of wind speeds in the case-study location. With marginally lower accuracy, the Rayleigh distribution represents this pattern but requires fewer estimated parameters.

Table 1. Weibull distribution parameters.

Parameter	Value	95% Confidence	Unit
c	7.70	[7.66, 7.75]	(m/s)
k	1.68	[1.67, 1.69]	(-)

Table 2. Rayleigh distribution parameters.

Parameter	Value	95% Confidence	Unit
b	5.71	[5.68, 5.73]	(m/s)

6 Conclusion

Cayo Carenas islet, situated at a tropical latitude, features high solar energy potential with a yearly average exceeding 5 peak sun hours per day. However, the probability of consecutive cloudy and rainy days, which may impact photovoltaic generation, must be accounted for.

The site's average wind energy potential is high (> 400 W/m^2), though not uniform year-round. Six months (January, February, May, October, November, December) show average performance, two (March, April) demonstrate outstanding values, and the remaining months exhibit low potential. Weibull and Rayleigh probability distributions were obtained, both satisfactorily modeling the random behavior of wind speed.

The characterization of Cayo Carenas' solar and wind potential confirms the islet's suitability for wind and photovoltaic generation systems. This assessment, along with the projected electrical demand survey for residential and commercial sectors, provides key inputs for designing a microgrid for the case-study community.

Acknowledgments. The research yielding the results presented in this work received funding from the International Funds and Projects Management Office, Havana, Cuba, under grant code PN211LH005-040, associated with the project Optimal Design and Predictive Control for Energy Management in Off-Grid Microgrids.

References

1. Zambroni de Souza, A.C., Castilla, M. (eds.): Microgrids Design and Implementation. Springer (2019)
2. Angelis-Dimakis, A., et al.: Methods and tools to evaluate the availability of renewable energy sources. Renew. Sustain. Energy Rev. **15**(2), 1182–1200 (2011). https://doi.org/10.1016/j.rser.2010.09.049
3. Ramachandra, T.V., Shruthi, B.V.: Spatial mapping of renewable energy potential. Renew. Sustain. Energy Rev. **11**(7), 1460–1480 (2007). https://doi.org/10.1016/j.rser.2005.12.002
4. Benedek, J., Sebestyén, T.-T., Bartók, B.: Evaluation of renewable energy sources in peripheral areas and renewable energy-based rural development. Renew. Sustain. Energy Rev. **90**, 516–535 (2018). https://doi.org/10.1016/j.rser.2018.03.020
5. Šúri, M., Huld, T.A., Dunlop, E.D., Ossenbrink, H.A.: Potential of solar electricity generation in the European Union member states and candidate countries. Sol. Energy **81**(10), 1295–1305 (2007). https://doi.org/10.1016/j.solener.2006.12.007
6. Martins, F.R., Pereira, E.B., Silva, S.A.B., Abreu, S.L., Colle, S.: Solar energy scenarios in Brazil, Part one: resource assessment. Energy Policy **36**(8), 2853–2864 (2008). https://doi.org/10.1016/j.enpol.2008.02.014
7. Makrides, G., Zinsser, B., Norton, M., Georghiou, G.E., Schubert, M., Werner, J.H.: Potential of photovoltaic systems in countries with high solar irradiation. Renew. Sustain. Energy Rev. **14**(2), 754–762 (2010). https://doi.org/10.1016/j.rser.2009.07.021
8. Ordóñez, G., Osma, G., Vergara, P., Rey, J.: Wind and solar energy potential assessment for development of renewables energies applications in Bucaramanga, Colombia. In: IOP Conference Series: Materials Science and Engineering, vol. 59, no. 1, p. 012004 (2014). https://doi.org/10.1088/1757-899X/59/1/012004
9. Ramgolam, Y.K., Soyjaudah, K.M.S.: Unveiling the solar resource potential for photovoltaic applications in Mauritius. Renew. Energy **77**, 94–100 (2015). https://doi.org/10.1016/j.renene.2014.12.011
10. Perpiña Castillo, C., Batista e Silva, F., Lavalle, C.: An assessment of the regional potential for solar power generation in EU-28. Energy Policy **88**, 86–99 (2016). https://doi.org/10.1016/j.enpol.2015.10.004
11. He, G., Kammen, D.M.: Where, when and how much solar is available? A provincial-scale solar resource assessment for China. Renew. Energy **85**, 74–82 (2016). https://doi.org/10.1016/j.renene.2015.06.027
12. Stolik Novygrod, D.: La energía FV: oportunidad y necesidad para Cuba. Econ. Desarrollo **152**, 69–86 (2014)
13. Gamez, M.R., Perez, A.V., Sera, A.S., Ronquillo, Z.M.: Renewable energy sources and local development. Int. J. Soc. Sci. Humanit. **1**(2), 10–19 (2017). https://doi.org/10.29332/ijssh.v1n2.31
14. González Lorente, Á., Hernández López, M., Martín Álvarez, F.J., Mendoza Jiménez, J.: Differences in electricity generation from renewable sources from similar environmental conditions: the cases of Spain and Cuba. Sustainability **12**(12) (2020). https://doi.org/10.3390/su12125190

15. Ilinca, A., McCarthy, E., Chaumel, J.-L., Rétiveau, J.-L.: Wind potential assessment of Quebec Province. Renew. Energy **28**(12), 1881–1897 (2003). https://doi.org/10.1016/S0960-1481(03)00072-7
16. Castellanos, F., Ramesar, V.I.: Characterization and estimation of wind energy resources using autoregressive modelling and probability density functions. Wind Eng. **30**(1), 1–14 (2006). https://doi.org/10.1260/030952406777641397
17. Migoya, E., Crespo, A., Jiménez, García, J., Manuel, F.: Wind energy resource assessment in Madrid region. Renew. Energy **32**(9), 1467–1483 (2007). https://doi.org/10.1016/j.renene.2006.06.015
18. Carta, J.A., Ramírez, P., Velázquez, S.: A review of wind speed probability distributions used in wind energy analysis: case studies in the Canary Islands. Renew. Sustain. Energy Rev. **13**(5), 933–955 (2009). https://doi.org/10.1016/j.rser.2008.05.005
19. Ucar, A., Balo, F.: Assessment of wind power potential for turbine installation in coastal areas of Turkey. Renew. Sustain. Energy Rev. **14**(7), 1901–1912 (2010). https://doi.org/10.1016/j.rser.2010.03.021
20. Ouammi, A., Dagdougui, H., Sacile, R., Mimet, A.: Monthly and seasonal assessment of wind energy characteristics at four monitored locations in Liguria region (Italy). Renew. Sustain. Energy Rev. **14**(7), 1959–1968 (2010). https://doi.org/10.1016/j.rser.2010.04.015
21. Ahmed, A.S.: Wind energy as a potential generation source at Ras Benas, Egypt. Renew. Sustain. Energy Rev. **14**(8), 2167–2173 (2010). https://doi.org/10.1016/j.rser.2010.03.006
22. de Araujo Lima, L., Bezerra Filho, C.R.: Wind energy assessment and wind farm simulation in Triunfo - Pernambuco. Brazil. Renew. Energy **35**(12), 2705–2713 (2010). https://doi.org/10.1016/j.renene.2010.04.019
23. Torres-Durán, A., Moreno-Figueredo, C.: Evaluación del potencial eólico en el Consejo Popular de Cojímar. Rev. Cubana Meteorol. **24**(3), 256–267 (2018)
24. Figueredo, C.M., Hernández, J.A.M., Ruiz, J.A., Janssens, A.: Dimensionamiento y selección de turbinas eólicas para el bombeo de agua en zonas pre-montañosas. Rev. Cubana Ingeniería **11**(3), 54–65 (2020)
25. Torres, Y.S., Ramos, P.R.: Techno-economic feasibility study of small wind turbines in Havana city. Rev. Cubana Ingeniería **12**(1), 47–57 (2021)
26. Toledo de la Cruz, D.: Carenas: una perla llena de trillos y curiosidades (+Galería) (2020). https://www.5septiembre.cu/carenas-una-perla-llena-de-trilloscuriosidades/
27. Barreras, R.: Cayo Carenas: la llave de la bahía de Cienfuegos (+Fotos) (2021). https://www.trabajadores.cu/20210129/cayo-carenas-la-llave-de-la-bahiade-cienfuegos-fotos/, https://www.trabajadores.cu/20210129/cayo-carenas-la-llave-de-la-bahia-de-cienfuegos-fotos/. Accessed 29 Jan 2021
28. Monteagudo Yanes, J.P., Vega Lara, B.G.: Entrevista a M.Sc. Iván García Gutierrez, Vice-Presidente de Cubasolar en Cienfuegos, residente de Cayo Carenas. Cienfuegos (2024)
29. Tanrioven, M.: Photovoltaic Systems Engineering for Students and Professionals: Solved Examples and Applications. CRC Press, Boca Raton (2023). https://books.google.com.cu/books?id=yEzUEAAAQBAJ
30. Ibrahim, S., Daut, I., Irwan, Y.M., Irwanto, M., Gomesh, N., Farhana, Z.: Linear regression model in estimating solar radiation in Perlis. Energy Procedia **18**, 1402–1412 (2012). https://doi.org/10.1016/j.egypro.2012.05.156. Terragreen 2012: Clean Energy Solutions for Sustainable Environment (CESSE)

31. Nelson, V., Starcher, K.: Wind Energy: Renewable Energy and the Environment. Energy and the environment. CRC Press, Boca Raton (2019). https://books.google.com.cu/books?id=ueQbtgEACAAJ
32. Ramírez, P., Carta, J.A.: Influence of the data sampling interval in the estimation of the parameters of the Weibull wind speed probability density distribution: a case study. Energy Convers. Manage. **46**(15), 2419–2438 (2005). https://doi.org/10.1016/j.enconman.2004.11.004
33. Rubiera, J.: Los Vientos De Cuaresma Y Otros Vientos (2021). https://www.excelenciascuba.com/generales/te-lo-conto-rubiera/los-vientos-decuaresma-y-otros-vientos

Predictive Models for Renewable Energy Generation in Microgrids. Cayo Carenas Case Study

David Antonio Fornet Cabrera(✉) and Boris Gabriel Vega Lara

Center for Energy and Environmental Studies, Faculty of Engineering, University of Cienfuegos, Cuatro Caminos, Carretera a Rodas Km. 4, Cienfuegos, Cienfuegos, Cuba
dafornet@ucf.edu.cu

Abstract. Intermittent renewable energy sources (RES) require forecasting of the power they will generate, which is particularly relevant in the operation of isolated microgrids. The objective of this work is to identify Autoregressive Integrated Moving Average (ARIMA) models for forecasting solar photovoltaic and wind generation in microgrid operations. The community of Cayo Carenas, an islet located in Cienfuegos Bay, Cuba, was used as a case study. The dataset was integrated by solar radiation, temperature, and wind speed measurements, recorded by a meteorological station near the islet. The technical parameters of the microgrid's solar and wind farms were used to estimate the power generated by each RES. Different ARIMA model configurations were evaluated using fit and complexity metrics. Forecast accuracy was evaluated for prediction horizons ranging from 5 to 20 min, with a 10-min horizon selected, yielding a mean absolute percentage error of 0.44% for the solar model and 1.51% for the wind model. The research highlights the importance of parameter selection and adapting the ARIMA model structure to the specific characteristics of each RES.

Keywords: Renewable Energy Source · Forecasting · ARIMA Model

1 Introduction

Microgrids are smaller-scale versions of power grids that can operate independently or synchronously to the main grid, which are emerging as smart grid subsystems capable of meeting the electricity needs of their own communities. Self-produced electricity has three possible flow options: to be consumed by the community, stored for later use, or exported to the main grid [1]. The particular characteristics of microgrids, with their capabilities to store and manage energy, allow intermittent and uncontrollable renewable energy sources (RES), such as solar PV and wind, to be properly managed to be converted into stable, continuous and reliable electric power supply resources [2].

A. Santos Fuentefria et al. (Eds.): JIFREEEMA 2025, LNNS 1933, pp. 126–136, 2026.
https://doi.org/10.1007/978-3-032-23028-7_12

Forecasting of renewable generation is fundamental to execute optimization tools, such as energy management systems, used for optimal scheduling of microgrid operation [3]. In conventional power systems, short-term forecasting is an effective method to control efficient system performance, contributing to operational decision making such as power dispatch scheduling, demand-side management, and generator performance monitoring [4]. Experts increasingly recognize the importance of very short-term forecasting in photovoltaic power generation, especially as integrating distributed energy resources (DERs) transforms power plant operations [5]. The integration of RES has advanced the prediction of the power delivered by such sources to a more challenging phase. The emergence of this complexity is due to nonlinearity and non-stationary load behavior [6].

In their beginnings, predictive models relied on statistical techniques, trying to synthesize in a more or less complex equation the underlying dependencies between the variable to be predicted and the factors that influence it [7]. In this case, the corresponding relationship between the forecast outcome and its determinants is clear, but it is necessary to understand the characteristics of the variable under study and to specify the type of model before forecasting [8]. The main advantage of statistical models is that they allow for managing the uncertainty associated with forecasts by quantifying the effect of each influencing variable on the total variability of renewable energy generation. In this group of statistical models, regression models and time series models stand out.

In recent studies, models based on machine learning have gained prominence. Their main advantage lies in their ability to identify more complex patterns and relationships in the data, as well as their ability to adapt and improve over time. They can make real-time predictions and automatically adjust their results based on sudden changes. This makes them ideal for analyzing nonlinearities that are difficult to express as equations in linear models [7]. It is not necessary to specify in advance the mapping relationship between the input variables and the forecasted variable, however, compared to the forecasting method based on mathematical equations, this method requires a large amount of historical data and the interpretation of the results is poor [8]. Their main disadvantage lies in the fact that they are black box models, which are not able to explain the underlying relationships between the dependent variable and the factors influencing it. Artificial neural networks and support vector machines belong to this group.

The most commonly used predictive models for forecasting of solar photovoltaic generation are the autoregressive integrated moving average model (ARIMA), the Kalman filter, the support vector machine, the least squares method, the artificial neural network, and hybrid models [3, 8]. In the case of wind generation forecasting, the most commonly used models and techniques are ARIMA models, Kalman filter, Markov chain, artificial neural network, support vector machine, and hybrid models [3, 8].

The ARIMA model, the Kalman filter and the artificial neural network are frequently used in forecasting the generation of both renewable energy sources. Given the strictly operational role that the prediction of solar and wind generation has in the real-time operation of microgrids, it is convenient to use the same type of predictive model in both cases in order to achieve certain homogeneity in the forecasts. The ARIMA model has the advantages that it is a dynamic model in time, it works with single variable data, and its structure is reduced to a linear equation of low complexity, which facilitates its implementation in the predictive controller algorithm. The neural network, although capable of providing more accurate forecasts, is a black box model with a high structural complexity that makes its implementation in the predictive controller algorithm very difficult, which is why it is selected as the predictive model to be used in this research. Several examples of case studies and applications of the ARIMA model and its variants in the forecasting of RES generation in microgrids and similar environments, such as small solar and wind farms, are presented below.

In the case of solar photovoltaic generation prediction, in [9], a solar generation prediction model is investigated and developed in a laboratory microgrid at the UCLA Smart Grid Energy Research Centre, based on historical solar radiation data and solar generation simulation using the System Advisor Model (SAM). The autoregressive moving average model (ARMA) was applied to predict solar generation in 2011, comparing its results with the persistence model. The ARMA model predicted solar generation with a mean absolute error of 0.0894 kW, lower than the error of 0.1213 kW achieved by the persistence model. In [10] they use seasonal ARIMA models (SARIMA) for generation forecasting of the solar parks of the academic block 1 and student hostel campus of IIT Gandhinagar, Gujarat, India. Solar generation data were measured at each park with a sampling frequency of 20 min, during the month of April 2017. The solar generation of both parks was predicted for the month of May 2017, yielding a mean absolute percentage error of 3.623% for academic block 1 and 3.395% for the student hostel.

In the case of wind power generation forecasting, in [11] several prediction models for wind power generation based on the ARIMA model are developed, using a meteorological database from the Brazilian National Environmental Data System (SONDA). Predictions are made for different time horizons, ranging from very short term (minutes) to long term (years), allowing a comprehensive assessment of wind energy availability. For the very short-term horizon, the hybrid ARIMA model with two neural networks showed a mean absolute error of 0.199 m/s, indicating good accuracy in predicting wind speed in very short time intervals. In [12] ARIMA models are identified for wind speed and direction at a specific site in Western Australia, with updates every minute over a seven-day period. The data were processed to obtain hourly averages. The results showed that the ARIMA model achieved a mean absolute percentage error of less than 5% for wind speed, indicating high prediction accuracy.

In [13], the frequency regulation in isolated microgrids considering the stochastic variability of wind and solar energy is worked out. Historical wind speed and solar radiation data obtained from Charanka Solar Park in Gujarat, India were used. SARIMA models were implemented for wind speed and solar radiation prediction, which achieved a mean absolute error of 4.704 W/m^2 for solar radiation and 0.090 m/s for wind speed, indicating good prediction accuracy. Table 1 summarizes the applications of ARIMA models and their variants discussed above.

Table 1. Summary of ARIMA model applications in renewable generation forecasting.

Source	Predictive Model	Sampling Rate	Forecast Horizon
[8]	ARMA	Hour to hour	1 h, 3 h
[9]	SARIMA	Every 20 min	20 min
[10]	ARIMA + RN1 + RN2	Minute to minute	20 min
[11]	ARIMA	Hour to hour	6 h
[12]	SARIMA	Hour to hour	24 h

The present work aims to identify Autoregressive Integrated Moving Average (ARIMA) models for forecasting solar photovoltaic and wind generation in microgrid operation. The contributions of this research are i) the use of high temporal resolution data (minute to minute) for the identification of ARIMA models and ii) the evaluation of forecast error at various very short-term forecast horizons (the next few minutes).

2 Methods and Materials

The Autoregressive Integrated Moving Average model (ARIMA) is a time series model that predicts future values from patterns identified in past data [14]. The ARIMA model has three components:

- AR-Autoregressive: Quantifies the linear relationship between the current value of the variable and its past values through the correlation between observations in different past periods. The number of past observations taken by the model is called p-order.
- I-Integrated: Represents the differentiation necessary to make the time series stationary, making its statistical properties constant over time. The number of differentiations to be applied to make the series stationary is called order d.
- MA-Moving Average: Captures the relationship between the current value of the variable and the forecast errors of past periods. The number of past errors taken by the model is called q-order.

As a whole, an ARIMA (p,d,q) model is represented in (1):

$$\nabla^d z_t = \phi_1 \nabla^d z_{t-1} + \phi_2 \nabla^d z_{t-2} + \ldots + \phi_p \nabla^d z_{t-p} + a_t - \theta_1 a_{t-1} - \theta_2 a_{t-2} - \ldots - \theta_q a_{t-q} \quad (1)$$

where

z_t is the current value of the variable;

a_t is the current error term;

ϕ_1, ϕ_2, y ϕ_p are the autoregressive coefficients of the model;

$\nabla^d z$ indicates that the time series is differenced;

θ_1, θ_2 y θ_q are the moving average coefficients of the model.

The construction of an ARIMA model and its subsequent use in obtaining forecasts is a process that has been widely addressed in the scientific literature [15, 16]. The stationarity of the time series was determined using the KPSS hypothesis test. Autoregressive

(p) and moving average (q) orders were obtained with the Akaike Information Criterion (AIC) analysis, calculated according to (2).

$$AIC = 2k - 2\ln(L) \tag{2}$$

where k is the number of parameters p, d and q of the model, and L is the likelihood function that measures the fit of the model to the observed time series data. A balance is sought between the model fit and its complexity, the higher the fit (L) the lower the AIC, however, as the model parameters (k) increase the AIC also increases. The AIC does not have an absolute value that can be interpreted by itself, but is used to compare different models, where a lower AIC is considered better than another with a higher AIC.

The autoregressive (ϕ_p) and moving average (θ_q) coefficients of the model were estimated using the maximum likelihood method. Evaluation of the forecasts was performed by running the obtained ARIMA models at 5, 10, 15, and 20-min forward prediction horizons, obtaining mean absolute error (MAE), root mean square error (RMSE), and mean absolute percentage error (MAPE) metrics (3–5) [17].

$$MAE = \frac{\sum_{i=1}^{n} |y'_i - y_i|}{n} \tag{3}$$

$$RMSE = \sqrt{\frac{1}{n}\sum_{i=1}^{n}(y'_i - y_i)^2} \tag{4}$$

$$MAPE = \frac{1}{n}\sum_{i=1}^{n}\frac{(y_i - y'_i)}{y_i} \tag{5}$$

where y_i are the real values of the variable and y'_i are the values predicted by the model.

For the development of the predictive models, there is a database provided by the Provincial Meteorological Centre of Cienfuegos. The data were taken every 10 min during 2011, by the Davis Station, an automatic weather station located near Cayo Carenas. Among the meteorological variables recorded were solar radiation and ambient temperature, used to estimate solar photovoltaic generation, and wind speed, used to estimate wind generation. As the sampling frequency of the microgrid controller is from minute to minute, the series of meteorological variables were interpolated. In the case of solar radiation and ambient temperature, measurements were taken for the period from April 5 to April 12, 2011 and a linear interpolation was applied. In the case of wind speed, measurements were taken for the period from February 5 to February 12, 2011 and a cubic spline interpolation was applied. For the three variables, 90% of the sample was used to adjust the models and the remaining 10% was used to validate the forecasts.

The procedure to obtain forecasts of the power to be generated differs for each RES. Solar photovoltaic generation is directly forecasted using an ARIMA model that works on the time series of generated power, so predictive models of the influential variables (solar radiation and temperature) are not developed to reduce the computational cost. The generated power series is estimated from (6) [13], taking as inputs the solar radiation and temperature series.

$$P_{PV} = \eta Sr[1 - 0.005(T_a + 25)] \tag{6}$$

where:

P_{PV} is the power generated by the solar photovoltaic system;

η is the conversion efficiency of the panels;

S is the area occupied by the solar panels;

r is the solar radiation;

T_a is the ambient temperature.

In the case of wind generation, the wind speed is first predicted using an ARIMA model, and then the power generated with the predicted speed is estimated using a polynomial regression model fitted to the wind turbine power curve, which shows the electrical power generated as a function of wind speed [18].

3 Results

3.1 Predictive Model for Solar Photovoltaic Power

The solar park of the Cayo Carenas microgrid occupies an area (S) of 98.01 m^2, and is made up of DSM 279 photovoltaic panels, with a conversion efficiency (η) of 10%. With these data and the radiation and temperature series, the time series of generated power (Fig. 1) is obtained from (6).

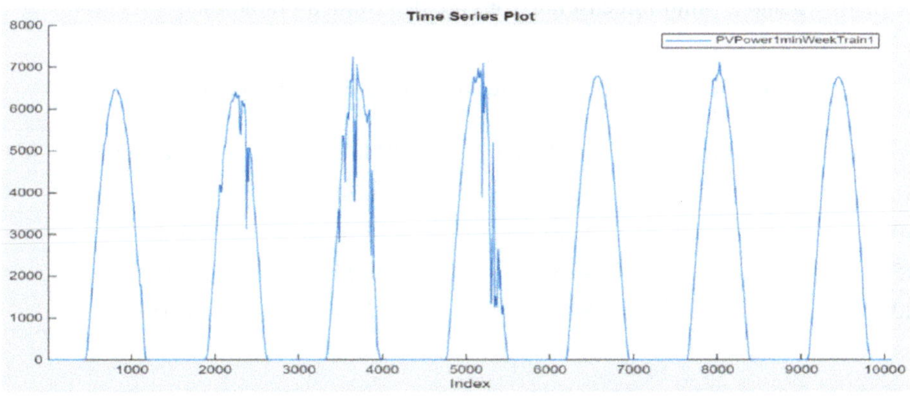

Fig. 1. Time series of PV power (5–12 April 2011).

To determine whether the series is stationary or not, the KPSS test is performed, where the rejection of the null hypothesis (H_0: The time series is stationary) with a P-value of 0.0172 allows us to conclude that the series is non-stationary with 95% confidence. The series is transformed by first order differencing. The KPSS test is performed again, this time on the differenced series, where the acceptance of the null hypothesis with a P-value of 0.1000 allows us to conclude that the series is stationary with 95% confidence. Therefore, the I component of the ARIMA model is of order d = 1.

Then, the structures of 20 ARIMA models with different combinations of orders (p,q) are evaluated and the Akaike Information Criterion value is determined (Fig. 2).

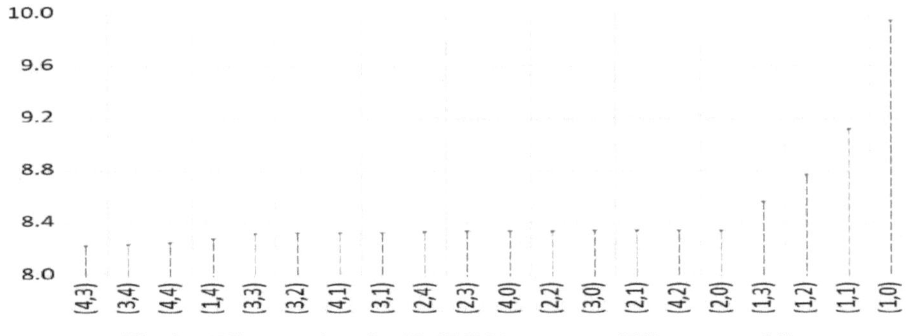

Fig. 2. AIC comparison for 20 ARIMA structures (PV power model).

Models (1,3), (1,2), (1,1) and (1,0) are excluded as candidates because they have values above 8.4. Of the remaining models, (2,0) is selected because it is the least complex. Once the AR coefficients have been estimated, the ARIMA model is defined by (7):

$$\nabla P_t = 0.9441 \nabla P_{t-1} - 0.0558 \nabla P_{t-2} + a_t \tag{7}$$

For the evaluation of the forecasts, the MAE, RMSE and MAPE metrics associated with the 5, 10, 15 and 20-min forecast horizons are determined (Table 2).

Table 2. Error metrics for different forecast horizons (PV power model).

Forecast Horizon (min)	MAE (W)	RMSE (W)	MAPE (%)
5	5.67	6.34	0.138
10	18.51	25.74	0.437
15	48.89	68.40	1.129
20	89.62	122.34	2.030

Errors increase significantly at the 15-min horizon, so it is recommended to set the forecast horizon between the next minute (P_{t+1}) and the next 10 min (P_{t+10}).

3.2 Predictive Model for Wind Power

The time series of wind speed (Fig. 3) is analyzed using the KPSS test, where the rejection of the null hypothesis with a P-value of 0.0100 allows concluding that the series is non-stationary. The series is transformed by first order differencing. The KPSS test is performed on the differenced series, with the acceptance of the null hypothesis with a P-value of 0.1000. Therefore, the I component order is d = 1.

To determine the p-order and q-order of the model, the structures of 20 ARIMA models are evaluated by calculating the value of the Akaike Information Criterion (Fig. 4).

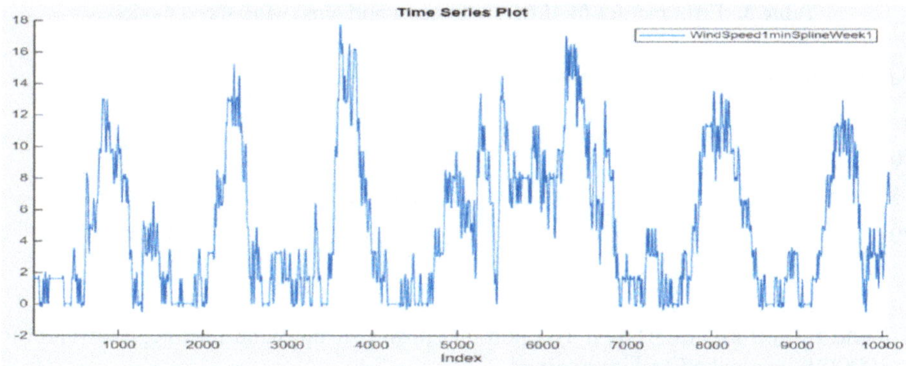

Fig. 3. Time series of wind speed (5–12 February 2011).

Fig. 4. AIC comparison for 20 ARIMA structures (wind speed model).

Models (4,0), (3,2), (4,1) and (4,2) are chosen as candidates, because they have values below -8. Model (4,0) is selected because it is the least complex of the four candidate models. Once the autoregressive coefficients are known, the equation of the ARIMA model is as presented in (8).

$$\nabla V_t = 3.0302\nabla V_{t-1} - 3.4213\nabla V_{t-2} + 1.6831\nabla V_{t-3} - 0.3007\nabla V_{t-4} + a_t \quad (8)$$

The MAE, RMSE and MAPE metrics associated with the following 5, 10, 15 and 20-min prediction horizons are shown in Table 3.

Errors increase significantly at the 15-min horizon, so it is recommended to set the forecast horizon to a value between the next minute (V_{t+1}) and the next 10 min (V_{t+10}).

The Cayo Carenas microgrid wind farm will consist of Enair E200L wind turbines, with start-up speed of 1.85 m/s, cut-off speed of 30 m/s, rated power of 10000 W. The relationship between wind speed and the power delivered by the wind turbine is modelled through a piecewise function, the intervals of which are explained below:

1. Wind from 0 to 1.85 m/s: The generated power is equal to zero because the wind speed is below the wind turbine start-up speed.
2. Wind from 1.85 m/s to 9 m/s: The power generated follows a polynomial function.

Table 3. Error metrics for different forecast horizons (wind speed model).

Forecast Horizon (min)	MAE (W)	RMSE (W)	MAPE (%)
5	0.034	0.038	0.498
10	0.118	0.190	1.511
15	0.546	0.872	7.453
20	1.126	1.624	18.721

3. Wind from 9 m/s to 30 m/s: The nominal power of the wind turbine, equivalent to 10000W, is reached and maintained.
4. Wind of 30 m/s or more: The generated power drops to zero because the wind turbine stops having exceeded its cut-off speed.

The polynomial function defining the power in the velocity range from 1.85 m/s to 9m/s is fitted as a polynomial of degree 2 with a coefficient of determination $R^2 = 0.9875$. It should be noted that a value as close to 1 in the coefficient of determination is considered over-fitting, however, in the context of this model the over-fitting is valid, because the polynomial obtained will only be used to predict the power generated by the Enair E200L wind turbine, and only in the speed interval over which the curve was fitted. For other wind speed ranges and other wind turbine models, it is necessary to perform new regressions with the new data.

Once the coefficients of the selected polynomial have been estimated, we proceed to present the piecewise function that defines the power generated from the wind speed in (9).

$$P(V_t)\begin{cases} 0 \leq V_t < 1.85; \; 0 \\ 1.85 \leq V_t < 9; \; 105.27V_t^2 + 395.44V_t - 1460.3 \\ 9 \leq V_t < 30; \quad 10000 \\ V_t \geq 30; \qquad 0 \end{cases} \tag{9}$$

4 Conclusions

In this work, ARIMA models were identified for forecasting solar photovoltaic and wind generation in microgrids, based on data series of meteorological variables and technical parameters of the microgrid under study. The models were trained with high resolution data and minute-by-minute sampling frequency, which makes them suitable for integration into optimization problems and predictive control algorithms for real-time operation of micro grids. The accuracy of the models was evaluated at prediction horizons of 5, 10, 15 and 20 min onwards, selecting the 10-min horizon as the most suitable for having a reasonable time range and forecast error.

Acknowledgments. The research that gives rise to the results presented in this paper received funding from the Office of International Projects and Funds Management under the code

PN211LH005-040, corresponding to the project "Optimal Design and Predictive Control for Energy Management in Off-Grid Microgrids".

References

1. Marinescu, A., Harris, C., Dusparic, I., Clarke, S., Cahill, V.: Residential electrical demand forecasting in very small scale: an evaluation of forecasting methods. In: 2nd International Workshop on Software Engineering Challenges for the Smart Grid co-located at ICSE (2013)
2. Zambroni de Souza, A., Castilla, M.: Microgrids Design and Implementation. Springer (2019)
3. Delfino, F., et al.: Microgrid Design and Operation. Toward Smart Energy in Cities. Artech House (2018)
4. Kondaiah, V.Y., Saravanan, B., Sanjeevikumar, P., Khan, B.: A review on short-term load forecasting models for micro-grid application. J. Eng. Pp. 665–689 (2022). https://doi.org/10.1049/tje2.12151
5. Fraga-Hurtado, I., et al.: Advanced Multivariate Models Incorporating Non-Climatic Exogenous Variables for Very Short-Term Photovoltaic Power Forecasting. Electricity **6**(2), (2025). https://doi.org/10.3390/electricity6020029
6. Azeem, A., Ismail, I., Muslim, J.S., Harindram, V.R.: Electrical load forecasting models for different generation modalities: a review. IEEE Open Access J. **9**, 142239–142263 (2021)
7. Hernández Callejo, L.: Microrredes eléctricas. Integración de generación renovable distribuida, almacenamiento distribuido e inteligencia. Grupo Editorial Garceta (2019)
8. Wang, H., et al.: A comprehensive review for wind, solar, and electrical load forecasting methods. Global Energy Interconnection **5**(1), 9–30 (2022). https://www.sciencedirect.com/journal/global-energy-interconnection
9. Huang, R., Huang, T., Gadh, R., Li, N.: Solar generation prediction using the ARMA model in a laboratory-level micro-grid. In: Proc. 2012 IEEE Third International Conference on Smart Grid Communications (SmartGridComm). IEEE, Tainan (2012). https://doi.org/10.1109/SmartGridComm.2012.6486039
10. Kushwaha, V., Pindoriya, N.: Very short-term solar PV generation forecast using SARIMA model: A case study. In: 2017 7th International Conference on Power Systems (ICPS), pp. 430–435. IEEE, India (2017). https://doi.org/10.1109/ICPES.2017.8387332
11. Barbosa de Alencar, D., et a.: Different Models for Forecasting Wind Power Generation: Case Study. Energies **10**(1976), (2017). https://doi.org/10.3390/en10121976
12. Yatiyana, E., Rajakaruna, S., Ghosh, A.: Wind speed and direction forecasting for wind power generation using ARIMA Model. In: 2017 Australasian Universities Power Engineering Conference (AUPEC). IEEE, Melbourne (2017). https://doi.org/10.1109/AUPEC.2017.8282494
13. Kumar, D., Mathur, H., Bhanot, S., Bansal, R.: Frequency regulation in islanded microgrid considering stochastic model of wind and PV. Int. Trans. Electr. Energ. Syst. e12049 (2019). https://doi.org/10.1002/2050-7038.12049
14. Iliya Gwaivangmin, B.: Examining different approaches for short-term load demand forecasting in microgrid management: a case study of a university in Nigeria. Academia Green Energy (1) (2024). https://doi.org/10.20935/AcadEnergy6220
15. Box, G., Jenkins, G., Reinsel, G., Ljung, G.: Time Series Analysis. Forecasting and Control, 5 th Edition. Wiley (2016)
16. Montgomery, D., Jennings, Ch., Kulahci, M.: Introduction to Time Series Analysis and Forecasting. Wiley (2008)

17. Mamun, A., Sohel, M., Mohammad, N., Sunny, M., Dipta, D., Hossain, E.: A Comprehensive Review of the Load Forecasting Techniques Using Single and Hybrid Predictive Models. IEEE Access **8**, 134911–134939 (2020). https://doi.org/10.1109/ACCESS.2020.3010702
18. Nelson, V., Starcher, K.: Introduction to Renewable Energy, 2nd edn., p. 2016. CRC Press, Taylor & Francis Group (2016)

The University Research Process Oriented Toward Sustainability: A Case Study of the Technical University of Manabí

Yadira del Pilar Chávez Loor[1](✉), María Dolores Chávez Loor[1], Neyfe Sablon Cossio[1], María Sonia Fleitas Triana[2], Daniel Alfonso Robaina[2], and Ariel Racet Valdés[2]

[1] Universidad Técnica de Manabí, Portoviejo, Ecuador
yadira.chavez@utm.edu.ec
[2] Universidad Tecnológica de La Habana "José Antonio Echeverría", La Habana, Cuba

Abstract. This study examines the research management process at the Technical University of Manabí (UTM), emphasizing its alignment with the Sustainable Development Goals (SDGs), particularly those related to the environment and sustainability. As a key territorial development actor, the university is responsible for producing knowledge that addresses social, economic, and ecological challenges. The main objective is to assess how the institutional research process contributes to the achievement of SDGs 2 (Zero Hunger), 4 (Quality Education), 5 (Gender Equality), 13 (Climate Action), 14 (Life Below Water), and 15 (Life on Land). A mixed-methods approach was employed, combining surveys and semi-structured interviews with 59 faculty researchers and 7 vice-deans in charge of Outreach, Research, and Graduate Studies. The findings highlight progress in scientific culture, evidenced by increased indexed publications, participation in research networks, and projects with a sustainability focus. Research lines are aligned with local environmental and social issues, and specialized centers foster collaboration among researchers, students, and communities. Despite regulatory advances, challenges persist in integrating SDGs transversally into research planning and management. Greater coordination among academic stakeholders is needed for effective implementation. The adoption of clear policies, capacity strengthening, continuous training, and active engagement with the productive and social sectors are essential for consolidating sustainable research management that effectively enhances quality of life and environment.

1 Introduction

The United Nations Sustainable Development Goals (SDGs), established in 2015 as part of the 2030 Agenda, provide a global framework for addressing humanity's most urgent challenges, including poverty eradication, climate action, and gender equality [1]. Within this context, higher education institutions play a strategic role as agents of change, with the capacity to generate knowledge, train critical professionals, and propose innovative

A. Santos Fuentefria et al. (Eds.): JIFREEEMA 2025, LNNS 1933, pp. 137–145, 2026.
https://doi.org/10.1007/978-3-032-23028-7_13

solutions that contribute to sustainable development [2]. However, effectively integrating the SDGs into university management remains a complex challenge, requiring not only political will but also clear institutional strategies and rigorous evaluation mechanisms [3].

The Technical University of Manabí (UTM), located in Ecuador, has actively committed to sustainability by aligning its research efforts with the SDGs, particularly those related to environmental and social dimensions [4]. Although universities are expected to contribute to achieving the SDGs by equipping students with the necessary skills and responses [5], significant gaps persist in aligning academic research with global sustainability goals [6]. Achieving the transformation envisioned by the SDGs demands strong participation from universities through their core functions: teaching, research, outreach, and institutional governance. These functions are crucial to generating and disseminating knowledge—essential actions for enabling change agents to transform society [7].

This study aims to analyze the contributions of UTM's research process to the fulfillment of the SDGs, with a particular focus on Goals 2 (Zero Hunger), 4 (Quality Education), 5 (Gender Equality), 13 (Climate Action), 14 (Life Below Water), and 15 (Life on Land), selected for their relevance to the regional context of Manabí. This coastal province faces critical challenges in terms of food security, deforestation, marine conservation, and social equity [8]. In addition, the study seeks to identify the factors that facilitate or hinder the alignment of university research with the SDGs, providing empirical evidence to support improved institutional policies. This makes UTM a relevant case for exploring how universities can respond to complex territorial contexts through sustainability-oriented research.

The relevance of this work is grounded in three key aspects. First, there is a gap in the literature: although studies have examined the relationship between universities and the SDGs [9], few focus on institutions in developing countries, where research resources are limited and socio-environmental challenges are more acute [10]. Second, UTM's strategic role as a public university, with a legal mandate to contribute to territorial development [11], positions it as an emblematic case for studying the feasibility of SDG implementation in budget-constrained academic environments. Third, there is an urgent need for empirical evidence to evaluate the actual impact of academic research on advancing the SDGs. The climate crisis and social inequalities demand immediate responses from academia [12], making it imperative to critically assess how university research contributes to sustainability.

This study is framed within contemporary theoretical debates on sustainability governance in higher education [13] and transformative science [14], which advocate for more collaborative, interdisciplinary, and action-oriented research models. It also draws on empirical evidence from successful cases of SDG integration in Latin American universities [15], while acknowledging critiques of the superficial "SDG-washing" of academia, where goals are referenced rhetorically without effective implementation [16].

In summary, this research not only aims to document UTM's progress in sustainability-focused research but also to propose a model for other institutions facing similar challenges. The findings will contribute to:

- Institutional policy: Designing guidelines to strengthen alignment between research and the SDGs.
- Teaching practices: Promoting the training of researchers committed to sustainability.
- International cooperation: Fostering alliances with academic networks working on the 2030 Agenda [17].

The article is structured into five sections: Methods, presenting a mixed-methods design; Results, integrating both quantitative and qualitative findings; Discussion, contrasting the evidence with existing literature; Conclusions, summarizing practical implications; and References, supporting the theoretical and methodological framework.

2 Methods

This study employed a mixed-methods design, integrating quantitative and qualitative approaches to analyze the contributions of the research process at the UTM to the achievement of the SDGs. The combination of methods allowed for the collection of objective data on scientometric and bibliometric indicators, as well as an in-depth exploration of the perceptions and experiences of key actors involved in research management.

A purposive sample was used, consisting of 59 active faculty researchers and 7 Vice-Deans responsible for Outreach, Research, and Graduate Studies (VIP) across various faculties at UTM. Participant selection was based on their active involvement in research projects aligned with the SDGs—specifically Goals 2 (Zero Hunger), 4 (Quality Education), 5 (Gender Equality), 13 (Climate Action), 14 (Life Below Water), and 15 (Life on Land).

The following instruments were used for data collection:

1. Structured surveys: Administered to faculty researchers. The survey included closed-ended questions and Likert-scale items aimed at assessing perceptions regarding SDG integration in research, resource availability, and institutional barriers.
2. Semi-structured interviews: Conducted with VIP Vice-Deans, focusing on identifying institutional strategies, research policies, and mechanisms for aligning research with the SDGs.

Quantitative data were processed using descriptive statistics (frequencies, percentages, and means) with SPSS v.25. Qualitative data from the interviews were analyzed through thematic coding, following the approach proposed in [18] which enabled the identification of emerging patterns and categories related to research governance and sustainability.

3 Results

The study reveals significant progress in aligning the research activities of the Technical University of Manabí with the Sustainable Development Goals, particularly the six goals selected as the focus of this analysis: Goals 2, 4, 5, 13, 14, and 15.

- SDG 2 – Zero Hunger: Approximately 27% of research projects address topics related to food security, sustainable agricultural production, or the sustainable management of resources. However, interviewees indicated that the outcomes of these studies have had limited influence on the design of local public policies.
- SDG 4 – Quality Education: Only 4% of research projects are linked to educational contexts, mostly within the university itself, indicating a low level of applicability in non-formal or non-university settings.
- SDG 5 – Gender Equality: While interviewees noted gender equity in the composition of research teams, they also pointed to ongoing gaps in resource allocation and in the visibility of gender-focused research outcomes.
- SDGs 13, 14, and 15 – Climate Action, Life Below Water, and Life on Land: Faculties such as Aquaculture and Marine Sciences, Agricultural Engineering, Agronomic Sciences, Agro-Sciences, and Veterinary Sciences, in collaboration with other academic units, are actively engaged in research on marine conservation, renewable energy, and biodiversity preservation. These efforts generate knowledge tailored to the ecological context of the coastal region of Manabí.

Academic production at UTM has shown both quantitative and qualitative growth. Between 2016 and 2023, the number of indexed publications increased from 12 to 2,072, representing an approximate cumulative growth of 17,17%. This exponential increase reflects a notable strengthening of research activities and an enhanced alignment with the SDGs.

In line with these results, 276 articles were identified as directly related to the SDGs, distributed as follows: SDG 12 (122 articles), SDG 14 (104), and SDG 13 (50). This distribution reveals a thematic concentration on environmental sustainability, resource management, and marine conservation.

In terms of research articulation, 74 active scientific collaboration networks were identified, of which approximately 20 are directly or indirectly related to climate change topics (SDG 13). These networks have facilitated inter-institutional cooperation, the exchange of specialized knowledge, and participation in international projects with a sustainability focus.

Regarding technology transfer, around 150 projects were documented that include some form of applied innovation and/or knowledge transfer strategies. Many of these initiatives incorporate sustainability components, particularly in the use of clean technologies, waste management, and agroecology.

Institutional perception was analyzed using data collected from surveys administered to faculty researchers (n = 59) and interviews with Vice-Deans for Outreach, Research, and Graduate Studies (n = 7). Key findings include:

- Enabling factors: 78% of faculty acknowledged that the SDGs guide their research lines, and 65% positively valued the existence of specialized centers promoting sustainability-focused projects.
- Identified barriers: 62% of respondents indicated the lack of specific funding for SDG-related projects as a major limitation. Additionally, 55% cited insufficient methodological training to evaluate the social impact of research.

From the perspective of the Vice-Deans, the results revealed the following:

- Institutional achievements: All interviewees (100%) highlighted that internal funding calls prioritize SDG-aligned topics, which has encouraged their inclusion in research proposals.
- Structural challenges: All respondents (100%) agreed that there are no integrated systems to measure the actual impact of research in terms of sustainability. Furthermore, 86% pointed to persistent fragmentation among faculties, which hinders interdisciplinary collaboration.

It was also found that only 40% of faculty members recognize the existence of clear institutional guidelines for integrating the SDGs into project design. Regarding interdisciplinarity, 83.38% of the projects are concentrated in STEM fields (science, technology, engineering, and mathematics), while only 16.62% are linked to the social sciences. This indicates an uneven distribution in the institutional research agenda.

3.1 Flagship Cases of Territorial Engagement

Several experiences of territorial intervention were documented through interdisciplinary outreach projects carried out by faculty and students from UTM, covering both STEM and social science domains. These initiatives include:

- "Productive and Healthy Manabí" Program (SDG 2): Aims to develop a comprehensive plan to support agricultural producers in achieving sustainable and environmentally responsible agro-livestock practices. The program addresses the indiscriminate use of agrochemicals that harm the environment and public health, while also seeking to reduce malnutrition in the province. It fosters collaboration among academia, public institutions, civil society, and grassroots organizations to meet the internationally agreed targets set by SDG 2: Zero Hunger.
- "Quality Education in Rural and Vulnerable Areas" Program (SDG 4): Seeks to improve education in rural zones and among vulnerable populations by implementing an integrated plan that strengthens educational quality and contributes to the endogenous development of local communities.
- "Healthy Marine-Coastal Ecosystems" Program (SDG 14): Aims to engage civil society in the understanding and protection of key species within coastal and marine ecosystems.
- "Take Care of the Earth and Save Your World" Program (SDG 15): Focuses on restoring ecosystems degraded by the loss of vegetation cover, with actions concentrated in prioritized areas.

Despite these promising initiatives, structural limitations remain regarding the institutionalization of sustainability at UTM. Although official documents explicitly reference the SDGs, project evaluation systems reveal only partial implementation. For instance, a low percentage of projects include sustainability indicators in their logical frameworks, which hampers effective monitoring. Furthermore, faculty training on the SDGs remains limited. The absence of mandatory training modules on sustainable development within internal capacity-building programs restricts the mainstreaming of this approach in the university's academic culture.

This situation reflects a case of "SDG-washing", where declarative commitments to the SDGs are not consistently translated into concrete and measurable practices.

Quantitative results were reported with 95% confidence intervals, while qualitative findings were supported by direct quotations from interviews, ensuring methodological rigor. This study directly addresses the objective of analyzing UTM's contributions to the SDGs, highlighting achievements, tensions, and opportunities for institutional policy reform, as discussed in the next section.

The results demonstrate that the UTM has made significant progress in aligning its research management with the Sustainable Development Goals, particularly in environmental and social domains (SDGs 2, 4, 5, 13, 14, and 15). These findings are consistent with the work of Leal Filho et al. (2018), which underscores the strategic role of universities as agents of change for sustainability—though structural challenges still constrain their real impact.

4 Discussion

The findings reflect a sustained growth in the scientific culture at the UTM, increasingly aligned with the SDGs. For instance, although 27% of projects address issues related to SDG 2 (Zero Hunger), interviewees reported a limited influence of these results on local decision-making processes. This disconnection between scientific output and practical application has been previously highlighted in the literature [16], who warn of the risk of superficial SDG-washing in academia—where SDGs are adopted rhetorically but lack mechanisms for real impact. The successful cases of research and outreach projects at UTM illustrate that community engagement is essential to move beyond the academic sphere, as proposed by the concept of transformative science [14].

Institutional willingness to align research with the SDGs is evident in internal funding calls and the establishment of specialized research centers, as emphasized by the Vice-Deans interviewed. However, 40% of faculty researchers reported the absence of clear guidelines to operationalize this alignment. Furthermore, all Vice-Deans (100%) acknowledged the lack of integrated systems to measure the social or environmental impact of projects—an issue that constitutes a critical barrier to evidence-based governance.

All interviewed stakeholders agreed that the institution prioritizes and supports SDG-aligned research projects, reflecting strong political will. Yet, the absence of standardized guidelines for project alignment (as reported by 40% of researchers), along with the total absence of impact assessment systems (as reported by 100% of Vice-Deans), reinforces concerns raised in [7] regarding the need for comprehensive regulatory frameworks. These are essential to address the observed fragmentation between STEM areas (which represent 83.38% of projects) and the social sciences (16.62%).

In summary, UTM has made significant progress in incorporating the SDGs into its research agenda. Nevertheless, structural challenges remain in terms of strategic planning, impact evaluation, and interdisciplinary collaboration. Advancing toward truly sustainable research requires moving from intention to institutionalization—through robust policies, effective monitoring mechanisms, and an academic culture genuinely committed to social transformation.

The findings from UTM reflect trends observed in other Latin American universities, where progress in aligning with the SDGs has been notable in environmental and productive areas, but structural challenges persist in cross-cutting integration, impact assessment, and interdisciplinary collaboration. Studies at institutions in Brazil, Mexico, and Colombia [19] show similar patterns of growth in indexed publications and collaboration networks, while also highlighting an overrepresentation of STEM projects to the detriment of the social sciences. This disparity limits the ability to address critical dimensions of the SDGs such as equity, governance, and community participation, which require qualitative and socio-critical approaches. Furthermore, the lack of specific funding for the SDGs—noted by 62% of researchers—affects different areas unevenly: while STEM projects often secure resources from technological or sectoral calls, social science initiatives face greater precariousness by depending on scarce internal funds or international cooperation. This asymmetry not only perpetuates disciplinary fragmentation but also weakens the comprehensive approach to sustainability, reinforcing the need for institutional policies that prioritize targeted funding and foster knowledge co-creation between STEM and the social sciences.

5 Conclusions

UTM has achieved significant progress in aligning its research management with the SDGs, particularly in environmental and social domains. Notable achievements include a 17.17% increase in scientific publications between 2019 and 2023, the establishment of 74 collaborative research networks, and meaningful community engagement initiatives.

However, the study also reveals structural tensions that hinder the consolidation of sustainability-oriented research governance. Key challenges include: the absence of cross-cutting policies that systematically integrate the SDGs, the limited incorporation of sustainability indicators in project frameworks, weak interdisciplinary articulation, and insufficient institutional capacity to evaluate real impact. Addressing these gaps requires a shift from rhetoric to institutional transformation. In this regard, the following priority actions are proposed:

Strengthen faculty development by integrating mandatory modules on sustainability and SDGs into internal training programs.
Embed sustainability indicators into the logical frameworks and evaluation systems of research projects.
Implement integrated monitoring and accountability systems based on empirical data and SDG-aligned metrics.
Promote interdisciplinarity by encouraging the inclusion of social sciences in the formulation of complex, systemic solutions.
Consolidate international partnerships, particularly with academic networks committed to the 2030 Agenda, to enhance institutional capacities and share best practices.

UTM stands at a key moment to establish itself as a regional leader in sustainability-driven research. Achieving this requires institutionalizing the progress made, addressing operational gaps, and embracing the SDGs not merely as a rhetorical framework but as a strategic guide to transform the university into a catalyst for inclusive, resilient, and knowledge-based territorial development.

References

1. United Nations: Transforming our world: the 2030 Agenda for Sustainable Development. United Nations, New York (2015)
2. Leal Filho, W., Shiel, C., Paço, A.: Sustainable development goals and sustainability teaching at universities: falling behind or getting ahead of the pack? J. Clean. Prod. **232**, 285–294 (2019). https://doi.org/10.1016/j.jclepro.2019.05.309
3. Purcell, W.M., Henriksen, H., Spengler, J.D.: Universities as the engine of transformational sustainability: toward delivering the sustainable development goals "Living labs" for sustainability. J. Environ. Stud. Sci. **9**, 1343–1357 (2019). https://doi.org/10.1108/IJSHE-02-2019-0103
4. Chávez, Y.P., Chávez, M.D., Sablon, N.: La investigación universitaria y los ODS: Caso UTM. Universidad Técnica de Manabí, Portoviejo (2023)
5. Sime Marques, R., Hernández, L.F., Delgado, M.J.: Los objetivos de desarrollo sostenible y su inclusión al currículo universitario: una revisión de literatura. Latam. Rev. Latinoam. Cienc. Soc. Humanid. **4**(2), 313–330 (2023). https://doi.org/10.56712/latam.v4i2.613
6. Tilbury, D.: Higher education for sustainability: a global overview of commitment and progress. Higher Educ. Sustain. **2**(1), 18–25 (2011)
7. Gutiérrez Mijares, A., Torres León, D., Gamboa, M.: Los Objetivos de Desarrollo Sostenible en la planificación estratégica de una universidad en Venezuela. Rev. Educ. Super. Soc. **36**(1), 335–357 (2024)
8. Secretaría Nacional de Planificación y Desarrollo – SENPLADES: Plan Nacional de Desarrollo 2017–2021. SENPLADES, Quito (2017)
9. Beynaghi, A., et al.: Future sustainability scenarios for universities: moving beyond the UN Decade of Education for Sustainable Development. J. Clean. Prod. **112**, 3462–3472 (2015). https://doi.org/10.1016/j.jclepro.2015.10.117
10. Tapia-Fonllem, C., Fraijo-Sing, B., Corral-Verdugo, V.: Sustainability in higher education institutions: its influence on the pro-sustainability orientation of Mexican students. SAGE Open **7**(1), 1–15 (2017). https://doi.org/10.1177/2158244016676295
11. Asamblea Nacional del Ecuador: Ley Orgánica de Educación Superior (LOES). Registro Oficial, Quito (2010)
12. Intergovernmental Panel on Climate Change – IPCC: Climate Change 2022: Impacts, Adaptation and Vulnerability. IPCC, Geneva (2023)
13. Davis, G., Goedegebuure, L.: Governance for sustainability in higher education. In: Singh, D., Stückelberger, C. (eds.) Ética en la educación superior: líderes impulsados por valores para el futuro, pp. 217–230. Globethics.net, Geneva (2021)
14. Horcea-Milcu, A., Abson, D.J., Apetrei, C.I., Fischer, J.: Transformative research for sustainability: characteristics, tensions, and moving forward. Glob. Sustain. **7**, e14 (2024). https://doi.org/10.1017/sus.2024.12
15. Vallaeys, F., de la Cruz, E., Sasia, P.M.: Responsabilidad social universitaria en América Latina. CINDA, Lima (2019)
16. Boström, M., et al.: Conditions for transformative learning for sustainable development: a theoretical review and approach. Sustainability **10**(12), 4479 (2018). https://doi.org/10.3390/su10124479
17. Hessen, D., Schmelkes, S.: La educación superior y los ODS: una síntesis basada en el informe del Grupo Global de Expertos Independientes sobre las Universidades y la Agenda 2030 (EGU2030). EGU2030, Paris (2022)

18. Braun, V., Clarke, V.: Using Thematic Analysis in Psychology. Qual. Res. Psychol. **3**(2), 77–101 (2006). https://doi.org/10.1191/1478088706qp063oa
19. Vallaeys F, Álvarez Rodríguez, J.: Hacia una definición latinoamericana de responsabilidad social universitaria. Aproximación a las preferencias conceptuales de los universitarios. Educación XX1 **22**(1), 93–116 (2019). https://doi.org/10.5944/educXX1.19442

Impact of the Corrosive Environment on the Support Structures of Photovoltaic Solar Panels

Laís Díaz Cordero(✉), Martha Mazorra Mestre, and Joel Morales Salas

Technological University of Havana Jose Antonio Echeverria, Havana, Cuba
laisdiaz297@gmail.com

Abstract. The implementation of photovoltaic solar parks has experienced significant growth in recent years and is expected to continue to do so in the coming years. This is why many scientists and manufacturers are increasingly developing more efficient and durable photovoltaic modules. To complement and achieve maximum sunlight capture, the panels must be positioned in an optimal orientation using support structures that are resistant to harsh weather conditions for long periods of time, leading to the onset of corrosion. The speed with which this process begins depends on both the internal characteristics of the material and the environment to which it is exposed.

The objective of this work is to evaluate general aspects of corrosion, the types of corrosion that may occur, the factors that accelerate this process, the classification of environments in Cuba, and possible alloys to use in these types of environments.

Keywords: Corrosive media · corrosion · photovoltaic solar panel support structures · metal coating

1 Introduction

In recent years, there has been exponential growth in the use of renewable energy worldwide, thanks to the awareness of many countries about the negative effects of the indiscriminate use of fossil fuels. This has allowed for a significant reduction in the emissions of polluting gases into the atmosphere, which have caused irreversible impacts on the planet and accelerated climate change.

Due to this great need to contribute to the care of the planet and the dire energy situation in Cuba due to the rising costs of fossil fuels globally, which makes access to these fuels impossible and leads to the search for alternative solutions, such as the use of renewable energy sources that reduce dependence on the use of fossil fuels.

Photovoltaic solar energy harnesses solar radiation to produce electricity immediately. The sun is the most abundant energy resource on the planet; it is estimated that in just one hour, all the energy consumed by humans for a year is received [1, 2], posing a challenge for humanity to achieve greater efficiency in capturing light energy. It constitutes a clean energy source, free from pollution and greenhouse gas emissions, and

A. Santos Fuentefria et al. (Eds.): JIFREEEMA 2025, LNNS 1933, pp. 146–154, 2026.
https://doi.org/10.1007/978-3-032-23028-7_14

is viable in the long term. For these reasons, photovoltaic solar energy is widely used worldwide as a sustainable alternative to conventional energy sources.

Given Cuba's geographical location in the tropics, photovoltaic solar energy is a sure bet for the optimal use of solar radiation in electrical energy, but it poses a challenge due to the climatic conditions characteristic of the tropical climate (hot and humid), a determining factor because it accelerates the rate of corrosion and puts at risk the stability and correct orientation of the support structures, preventing them from having a useful life for a time similar to that of the photovoltaic solar panel.

1.1 Operating Principle

Photovoltaic solar panels are composed of solar cells, which are responsible for converting solar energy into electricity. This process occurs in the internal layers below the reflective layer, where there are two layers of silicon, the upper one is a negative conductor, made of silicon with a mixture of phosphorus. The lower one is a positive conductor and is made with a layer of silicon with a mixture of boron. When the photon hits the silicon atom, the electrons begin to circulate between the plates, generating the movement of an electrical charge and a potential difference is generated that orders the movements of electrons and the appearance of an electrical voltage. The inverter is another essential component in the photovoltaic installation because it is responsible for converting the direct current received from the photovoltaic solar panels into alternating current, for later consumption. The transformers are responsible for raising the voltage of the alternating current generated by the inverters for transport to the medium voltage electrical grid [2–4]. Another fundamental component is the charge controller. This is responsible for regulating the flow of energy from the photovoltaic solar panels to the batteries, controlling both the intensity and voltage they receive, allowing charging to occur under optimal conditions and preventing battery damage. Finally, the batteries, which are the devices responsible for storing the energy captured and transformed by the photovoltaic system, provide energy availability at night or during intermittent periods when sunlight is minimal [5].

1.2 The Support Structure for Photovoltaic Solar Panels

The maximum efficiency of solar panels depends largely on an adequate inclination and orientation, thus allowing to optimize the amount of solar energy that can be captured and converted into electricity. By placing the panels at an appropriate angle, it is ensured that they are receiving the maximum amount of solar radiation possible. For the aforementioned reasons, the support structures of the photovoltaic solar system are of extreme importance because they are responsible for providing the solar panels with the inclination angle that guarantees the performance of a solar installation. They must present high resistance in extreme weather conditions in an adequate and stable manner for years (approximately 25 years). Therefore, a study is needed to guarantee the optimal performance of the materials used in the construction of these structures, avoiding premature failures that impede the performance of the photovoltaic modules [6]. These structures can be fixed or mobile. The fixed ones provide a stable inclination, defined by the geographical area to maximize the efficiency of the installation. Mobile solar panels are designed to increase the incidence of solar radiation from one to two axes (north-south, east-west) by directing the modules using solar tracking [7].

Currently, there is a wide variety of support structures for solar panels that can be adapted to different needs and environments, allowing installations ranging from ground-anchored installations to installations on roofs or inclined metal roofs, to installations on vertical surfaces. The choice of the type of structure to use depends on the type of roof where the photovoltaic system is to be installed and the climatic conditions of the site.

One of these is coplanar structures (Fig. 1), which are placed on inclined roofs. The positive aspect of these structures is that they do not visually affect the landscape as they are integrated into the design of the home or facility where they are installed.

Fig. 1. Support structure for photovoltaic solar panel on a sloping roof

Another type of structure for solar panels on flat or ground-level roofs, this type of support structure is the most widely used in Cuba, both in photovoltaic parks and in homes, due to the predominant architectural style in the country. A disadvantage is their susceptibility to wind effects due to their lack of rigidity (Fig. 2).

Fig. 2. Solar panel support structure on a flat roof

Vertical supports are anchored to facades or walls and must be resistant to strong winds (Fig. 3).

Fig. 3. Photovoltaic solar panel support structure on vertical support

Mobile solar supports, they have mobile axes and sensors that change according to the solar movement Fig. 4.

Fig. 4. Mobile photovoltaic solar panel support structure

1.3 Impacts on Support Structures

Solar panel support structures are exposed to climatic events such as rain, humidity, dew, and fog throughout the year, so they will be exposed to high relative humidity, which favors corrosive processes. Furthermore, due to Cuba's geographical characteristics, these structures will be impacted by weather events such as cold fronts and tropical cyclones, so resistance to strong winds is a factor to consider, as it is difficult to assemble and disassemble large photovoltaic parks for maintenance or to avoid damage from the impact of weather events.

Impact of Corrosion

Corrosion is defined as the spontaneous process, the reverse of metallurgy, causing metallic materials to return to their natural state or a similar form to that in which they were found in nature, due to chemical or electrochemical attacks resulting from interaction with the surrounding environment [8]. Therefore, the rate at which this process occurs depends on the type of material and the conditions to which it will be exposed. Hence the importance of evaluating the material of the support structures for photovoltaic solar panels and its behavior over long periods of time in the specific location where it will be installed.

The corrosion process is classified into two large groups: one is chemical corrosion, which occurs in metals in contact with dry gases and at high temperatures, and the second is electrochemical corrosion, which is the type of corrosion that solar panel support structures may exhibit [8, 9].

Electrochemical Corrosion Process

The electrochemical corrosion process is the most common in nature since it occurs at room temperature and occurs as a spontaneous process through the action of three elements, which are essential for the process to take place. These three factors are: the existence of an anodic zone (which is simply the area that suffers corrosion), the cathodic zone (which could be the same metal with heterogeneous interfaces, or different metals), and an electrolyte, which is often condensed water from the atmosphere (for this to occur, the relative humidity must be 70%). The combination of these three factors produces the transfer of electrons from one metal to another, generating an electric current from the metallic anode to the metallic cathode, such that the anode is oxidized (transfers electrons), while the cathode is reduced (gains electrons) [10, 11].

All electric current is generated by a potential difference between two electrodes that form the anode and the cathode, which, once electrically connected, generate an electric current caused by the flow of electrons from the anode to the cathode. The intensity of the current will depend on the medium through which these electrons circulate and, obviously, on the existing potential difference. This phenomenon, as previously mentioned, is what causes metal corrosion in most cases, especially atmospheric corrosion.

Thus, for electrochemical corrosion to occur, an electrochemical potential must exist that is capable of activating the flow of electrons necessary for the process to occur. Differences in electrochemical potential are the origin of some regions acting anodically and others acting cathodically. In this sense, the electrochemical potential can be caused by:

- Metals of different nature and with significant electrochemical potential.
- In the same metal, due to the heterogeneities caused by different phases within it, which in turn generate the electrochemical potential necessary for the process to occur.

It depends on both internal factors, such as the passivation capacity of the metal used in the environment where it will be exposed, the presence of impurities, or the surface condition of the metal, and external factors, such as the aggressiveness of the environment where the support structures for the solar panels are to be placed.

Corrosion in support structures for photovoltaic solar panels is characterized by non-homogenized or non-uniform corrosion, without mechanical effects. Within this group, specifically pitting corrosion, occurs when the protective oxide layer on the surface breaks and comes into contact with a corrosive aqueous solution, making the part susceptible to electron loss and causing small spots to appear on the surface, thus accelerating the corrosion process.

The support structure and the edge of the solar panel are often made of different materials, which can produce galvanic corrosion effects. This type of corrosion occurs due to the presence of two materials and an electrolyte, forming an electrochemical cell where the material with the lower potential corrodes. It is important to note that in this process, only one of the two metals corrodes, and the other is virtually undamaged. The one that corrodes is called the active metal, while the one that is not damaged is called the noble metal. The greater the potential difference, the greater the possibility of galvanic corrosion occurring. To avoid this type of corrosion in solar panel support structures, it is necessary to use metals that do not have a very large potential difference.

Typically, in photovoltaic solar panel support structures, the starting point for this type of corrosion is the use of screws made of different materials than the support structure. Drilling holes exposes the base material. In the design of the support structure, there may be inclination angles that allow water to accumulate, or due to an incorrect selection of the material according to the environment where the support structure for the photovoltaic solar panel will be installed.

In addition to the aforementioned causes, there are external factors that accelerate the rate at which the corrosion process occurs. These factors are relative humidity (above 70%) and high temperatures. These two factors have a marked influence on the increase in the rate of chemical reactions that result in the deterioration of metals. These factors also contribute to other factors such as solar radiation, an important factor since the support structures will be exposed to strong solar radiation throughout the day. While these factors are determining factors, there are others that multiply the rate of corrosion. These are atmospheric pollutants such as sulfur dioxide (SO_2), nitrogen oxide (NO_X), atmospheric particles of different sizes or chemical compounds, and the presence of chlorides or marine spray. Taking into account all these external factors and complying with the Standard of the European Committee for Standardization, estimates of the corrosion rate are offered for different environments, classified by corrosivity category (C1-CX), as shown in Table 1:

Table 1. Atmospheric corrosivity categories according to ISO 9223 standards [9, 12].

Category	Corrosiveness	Characteristics
C1	Very Low	Low relative humidity, negligible contamination
C2	Low	Variable temperature and relative humidity. Low frequency of condensation and low contamination
C3	Medium	Moderate frequency of condensation and moderate contamination from production processes. Medium SO_2 contamination
C4	High	Areas with high frequency of condensation and high contamination from production processes. High SO2 contamination. Some tropical areas are in this category
C5	Very High	Very high frequency of condensation and/or high contamination from the production process, with significant chloride contamination
CX	Extreme	Almost permanent condensation or extended periods of exposure to extreme humidity and/or high contamination from the production process. Occurs in humid tropical areas with penetration of external contamination, including airborne chlorides and corrosion-promoting particles

Considering that Cuba is characterized by a warm, humid climate located geographically in the tropical zone, it is considered to be in the high (C4), very high (C5), and extreme (CX) corrosivity categories, depending on the distance from the coastline and the atmospheric classification of the exact location where the photovoltaic solar panel support structure is to be installed [13].

In order to significantly extend the lifespan of solar panel support structures, it is necessary to apply a metallic coating (hot-dip galvanized) with alloys that are corrosion-resistant for long periods of time. It is worth noting that a zinc-aluminum alloy resistant to extreme conditions requires a coating greater than 100 µm thick.

Figure 5 shows the durability of galvanized steel taking into account the different categories of aggressiveness in the different environments (C1 - CX), relating the coating life in years and the coating thickness [14].

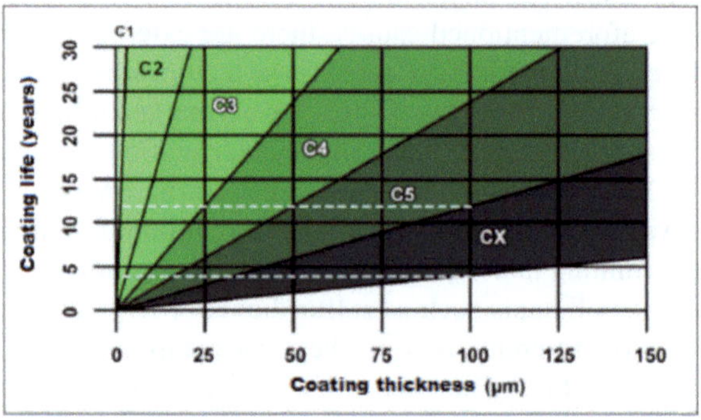

Fig. 5. Durability of Galvanized Environments (C1–CX) [14]

Zinc–Aluminum–Magnesium (Zn–Al–Mg) Alloys

The corrosion resistance of hot-dip coatings can be improved by increasing the levels of aluminum alloy addition. Good examples of these coatings are Galfan Zn + Al (5 wt%) and Galvalume Al (5% by weight) and Galvalume Al (55% by weight) + Zn (43.5% by weight) + Si (1.5% by weight). These coatings offer higher levels of corrosion resistance relative to pure zinc because they combine the protection of zinc with a durable physical barrier of aluminum oxide [15].

Ternary coatings containing zinc, aluminum, and magnesium (Zn-Al-Mg coatings) have also been developed and are known as "ZM" coatings. Commercial compositions for these hot-dip coatings range from 1 to 11% in weight of aluminum and 1–6% in weight of magnesium, with the remainder of the coating being zinc[9].

Among the (Zn-Al-Mg) alloys, Magnelis® stands out, an exceptional metallic coating that represents a step forward in corrosion protection. It has a wide range of applications thanks to its unique chemical composition. It is manufactured on a classic hot-dip galvanizing line, with a bath that has a special chemical composition based on 93.5% zinc, 3.5% aluminum, and 3% magnesium [9, 16].

Magnelis® is a steel specifically developed by ArcelorMittal. The first studies began in 2010 by R&D teams outside Avilés, although Avilés played an important role in the product's launch in 2017. It is currently manufactured by ArcelorMittal in Ghent (Belgium), Bremen (Germany), Ostrada (Czech Republic), and Asturias (Spain). In recent years, its use has shown significant growth due to its high corrosion resistance, with a 25-year warranty depending on the environment where it will be installed, with high resistance even in the most hostile environments, which has placed it in the European region as a leading product for the manufacture of solar panel support structures, due to its self-protection property on cut edges and perforations [17].

However, this new material is not widely used in Cuba despite its potential in several applications in several European countries. It is applicable to a wide range of steel grades, from cold-forming and deep-drawing steels to structural steels and high-strength, low-alloy steels.

The steel thickness can range from 0.45 to 6 mm, while the coating can vary from 5 to 35 µm per side. This metallic coating forms a very dense, stable, and durable protective layer, acting as a corrosion barrier and preventing the underlying steel from coming into contact with the surrounding environment.

The dense layer formed is present even in highly deformed areas, so that both flat and deformed areas enjoy the same protection across the entire surface. A key advantage over other metal coatings is the formation of this protective layer on edges, welds, perforations, and scratches. Should red rust appear in these areas, it will gradually be covered by the Magnelis® film, providing almost perfect protection for periods of more than 20 years.

Thanks to its highly resistant, adherent layer, it offers the possibility of deformation using various methods such as bending, stamping, and profiling. The manufacturer reports improved results in terms of product malleability and protection of work tools. Friction tests conducted by the manufacturer demonstrate that it offers better results than hot-dip steel. It can be processed without the need for lubricants, something that seems impossible with galvanized steels, and it offers improved weldability following the same procedure as zinc-coated parts, using the same consumables and welding guidelines. In addition, a subsequent painting process offers greater corrosion resistance.

2 Conclusions

The different types of solar panels and the types of corrosion that may occur are highlighted, highlighting galvanic corrosion due to the interaction of two different metals with a significant potential difference or the interaction of heterogeneous interfaces of the same metal or alloy.

The importance of characterizing materials before implementing their use is evident, allowing for the study and prediction of their behavior depending on the aggressiveness of the environment where they are to be installed. Cuba is categorized as highly corrosive (C4), very highly corrosive (C5), and extreme (CX) environments.

Hot-dip coatings are identified, such as (Zn-Al) alloys, highlighting the need for coating thicknesses greater than 100 µm, and the Zn-Al-Mg Magnelis® alloy, which allows for reduced coating thicknesses while achieving similar or superior results for the same period of time.

References

1. Moreno Figueredo, C., et al.: Fuentes renovables de energía. Tecnología y aplicaciones, pp. 1 (2021)

2. Balsinde Rodriguez, C.: Análisis comparativo de recubrimientos metálicos para estructura soporte de paneles solares fotovoltaicos en Cuba. Trabajo de Diploma, Universidad Tecnológica de La Habana José Antonio Echeverria CUJAE, Facultad de Mecánica (2024)
3. Iturralde Carrera, L.A., Castro Perdomo, N.A., Bambi, V.E.: La energía solar fotovoltaica y sus particularidades en Cuba. Revista Mapa **8**(23), 133–150 (2021)
4. Francisco Andrea, E.F.: Modelización y estudio de sistemas fotovoltaicos bifaciales Tesis de fin de grado. Escuela Superior de Ingeniería y Tecnología (2022)
5. Blog Enlight México: Baterias para paneles solares. Qué son, para qué sirven y cuanto cuestan (2021)
6. Moreno Figueredo, C.: Fuentes renovables de energía. Tecnología y aplicaciones. Libro de textos, Editorial Cubasolar (2021)
7. Autosolar, Estructura paneles solares/ estructura para placas solares. Autosolar
8. Salazar Jimenez, J.A.: Introducción al fenómeno de la corrosión: tipos, factores que influyen y control para la protección de materiales. Tecnología en marcha, vol. 28, no. 3 (2015) [consultado el 23 de agosto de 2023]. Available from: http://www.scielo.sa.cr/pdf/tem/v28n3/0379-3982-tem-28-03-00127.pdf
9. Chenoll Mora, E., Cloquell Ballester, V.A.: Analyss of metallic coatings based in Zinc – Aluminium – Magnesium alloys, in terms of reformance and long – term corrosion. Case study: Electrical cable trays selection in proyect design,. Tesis doctoral (2021) [Consultado el 6 de mayo de 2023], 147. Available from: http://riunet.upv.es/bitstream/handle/10251/167418/Chenoll%20-%20Analyss%20of%20metallc%20coatngs%20basedn%20znc-alumnum-magnesum%20alloys%2Cn%20terms%20of%20performance%20an...pdf?sequence=1
10. Chunga Santos, G.E., Gaibor Encalada, F.D.: Análisis de la corrosión en aleaciones metálicas en Salinas (Ecuador) para determinar su empleabilidad en soportes para paneles solares. Trabajo de Diploma, Escuela superior politecnica del litoral (2022)
11. Martinez Pérez, F.: Corrosión tipos. Prevención. Rev Cie TécArg, vol. 32, no. 2 (2023). San José de las Lajas abr – jun. Publicación electrónica 2023. Available from: http://scielo.sld.cu/scielo.php?pid=S2071-00542023000200010&script=sci_arttext&tlng=es
12. Isisarri Alzate, Iker: Estudios de la resistencia a la corrosión de recubrimientos galvanizados y aleaciones de zinc. Universidad pública de Navarra (2022) [Consultado el 10 de agosto de 2023]. Available from: http://academica-e.unavarra.es/handle/2454/44275
13. Viña-Rodriguez, J., Castañeda Valdés, A., Valdes Clemente, C.: Corrosión atmosférica. Conceptos básicos y experiencias obtenidas en el clima tropical coostero de Cuba. CENIC Ciencias Químicas **52**(2), 121 (2021)
14. Industrias H7: Galvanizado por inmersión en caliente. Notas informativas Industrias H7 (2021)
15. Eaves David, R.: A study of novel filiform corrosion phenomena on hot dip organically coated Zn–Al– Mg steel. Thesis Swansea University (2013) [Consultado el 25 de Agosto de 2023]. Available from: http://cronfa.swan.ac.uk/record/cronfa43198
16. ArcelorMittal: Magnelis®-Aceros con recubrimientos de zinc – aluminio –magnesio. Revista Arcelor Mittal (2020) [Consultado el 11 de octubre de 2023]. Available from: http://contructalia.arcelormittal.com/es/productos/magnelis
17. ArcelorMittal: Magnelis® Desición estratégica. Magnelis® book (2019) Consultado el 15 de septiembre de 2023]. Available from: http://acerovial.com/documentos/SECCION%203/magnelis-book-am.pdf

Frequency Behavior of the Cuban Electrical System for Different Scenarios with the Introduction of 1200 MW Photovoltaic

Alfredo López Valdés[2], Antonio Martínez García[1], Miriam Vilaragut Llanes[1](✉), Miguel Castro Fernández[1], Mercedes Canasí Piñeiro[1], and Susana Leyva Condef[2]

[1] Technological University of Havana José A. Echeverría, CUJAE, Havana, Cuba
miriamv@electrica.cujae.edu.cu
[2] Electric Union, Havana, Cuba

Abstract. In Cuba, a change in the energy matrix is taking place with the incorporation of 1200 MW of photovoltaic power in parks of 21.8 MW by December 2025. In the present work, the behavior of grid-following and grid-forming inverters in primary frequency control is initially analyzed. As the PFVs have grid-follower inverters, the modeling and simulation is performed using a concentrated model to compare the frequency behavior when the control is performed by conventional units and tracker type inverters, taking into account the estimated variability of solar photovoltaic generation for December 2025 and a utilization of 80%. The results obtained by simulating the frequency behavior in December 2025 with the incorporation of about 1200 MW of solar photovoltaic generation without energy storage and considering the incorporation of 100 MW and 100 MWh of BESS battery energy storage systems used for primary frequency control with grid-forming converters are also presented.

Keywords: frequency behavior · renewable sources · BESS systems

1 Introduction

The introduction of renewable energy sources (RES) in the Electric Power Systems has had an important increase in the last years. This is related to the need to reduce environmental pollution caused by the use of fossil fuels derived from petroleum, which is causing an increase in the planet's temperature, the cause of environmental catastrophes that we are already beginning to experience.

Our country is experiencing a critical situation in the supply of electricity to the industrial, residential and service sectors, due to a generation structure composed of conventional units that use fossil fuels, Cuban crude oil and gas from oil exploration, which have been in operation for more than 35 years and in recent years have had their maintenance periods extended for economic reasons. Additionally, no investments have been made to increase installed generation, which currently causes deficits in the generation-load balance.

A. Santos Fuentefria et al. (Eds.): JIFREEEMA 2025, LNNS 1933, pp. 155–166, 2026.
https://doi.org/10.1007/978-3-032-23028-7_15

The situation of the country's electricity system at this time is critical, in addition to the lack of generating units that have the necessary technical conditions to be able to carry out an acceptable frequency and voltage control, which affects the quality of supply, as well as the safety of the system's operation.

In association with the above-mentioned conditions, the country has adopted strategic decisions aimed at resolving this situation in the long term, which are basically based on the following:

- Increase the total generation capacity of the system in the shortest possible time by installing solar photovoltaic and wind farms.
- Use BESS battery energy storage systems, with two objectives, primary frequency control and load shifting to the peak load period.
- Develop a general maintenance plan for the thermal units in service to recover, as far as possible, their nominal values and restore to some of the conventional units the possibility of being able to perform primary frequency control.
- It is also intended to install, before 2030, two new 200 MW conventional units that work with Cuban crude oil.

As part of this strategy, we are working intensively to achieve the installation, before December 2025, of 55 photovoltaic solar parks of 21.8 MW each, which represents an increase of just over 1,200 MW in one year. Also included is the installation of 100 MW / 100 MWh in BESS systems that should carry the fundamental weight in the primary control of the frequency of the system.

This paper presents the results of the simulations carried out for two essential scenarios: the behavior of the system with the incorporation of the 1200 MW photovoltaic without and with the installation of BESS systems and the comparison between both.

2 Behavior of Tracker and Shaper Inverters in Primary Frequency Control

To address the aforementioned situation, it is necessary to study the behavior of the existing forming converters in the energy storage systems that perform primary control of the frequency and of the grid follower inverters in the photovoltaic solar parks that, according to the Cuban grid code, must perform primary control of the frequency in the event that it rises above the nominal frequency of the system.

In [1] an interesting comparison is developed in the transient behavior of the frequency when the control is developed by conventional units and by inverters type trackers and formers. For this purpose, concentrated models of the system are used. As a result of this comparison it is concluded that the behavior of these inverters in the primary control can be considered similar to a conventional unit, only that the inertia and damping will be in the first case dependent on the delay time corresponding to the measurement and control droop as shown in Fig. 1.

where: Mp,inv = Droop of the control and TM = Measurement delay time.

In [1] it is pointed out that the longer measurement delay time in tracker inverters is due to the need for the PLL to synchronize with the system, which is not the case for

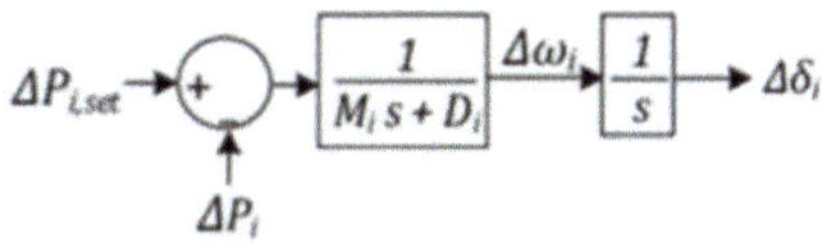

$$M_i \triangleq T_M/M_{p,inv} \quad D_i \triangleq 1/M_{p,inv}$$

GFL parameters		GFM Parameters
$M_{p,inv} = 0.05\omega_s/pu$		$M_{p,inv} = 0.05\omega_s/pu$
$T_L = 0.5s$		$T_M = 7.96\ ms$

Fig. 1. Equivalent diagram of frequency control in tracker and shaper inverters and typical values of the parameters in the equivalent system blocks

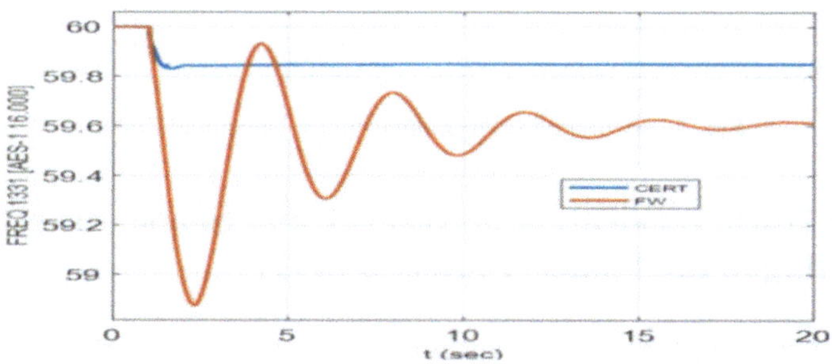

Fig. 2. Response to a step-up in control in the case of follower and shaper inverters. Taken from Reference [1].

shaper inverters. This is the reason why the frequency response in the first case behaves with much less damping, as can be seen in Fig. 2.

The scenario presented for December 2025 with a significant % of solar photovoltaic generation, which have inverter trackers, in which they must participate in frequency control corresponds to a scenario with low damping in which significant frequency deviations may occur. BESS systems are equipped with shaper inverters which, due to their fast response, introduce damping into the system. However, these will not be available until December 2025.

3 Variability of Solar Photovoltaic Generation Estimated for December 2025

Any study on the introduction of solar photovoltaic generation requires the consideration of its variability due to the passage of low clouds over the panels [3–7]. These variations can reach more than 50% of their nominal values in 1 s intervals [7, 8]. In recent studies carried out in the Cuban SEP, this variability has been measured in 5-min windows resulting in values in the order of 17–20% of the installed capacity [9–11].

Based on the solar generation patterns of 6 of the newly installed 21.8 MW solar parks and applying the methodology proposed in [10, 11], this generation is estimated resulting in the resulting pattern from 10 am to around 7 pm, shown in Fig. 3.

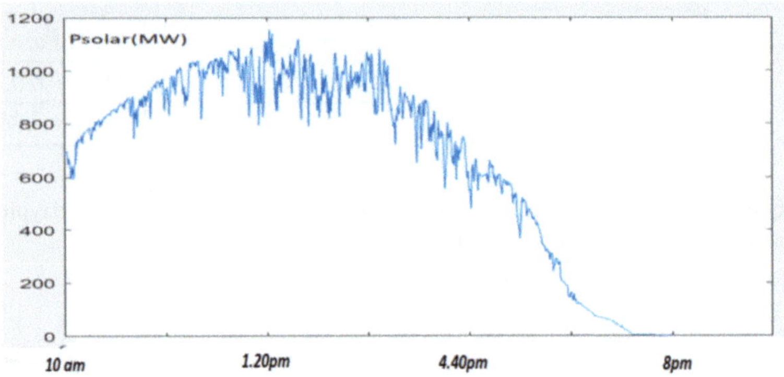

Fig. 3. Estimated solar generation by December 2025.

4 Model Work Verification of Tracker and Shaper Inverters According to the Models Proposed in [1]

To check the correct operation of the inverter tracker and shaper models, a concentrated model is used representing a 50 MW BESS system whose frequency control system is executed by inverter trackers and two shaper models, one with droop control and the other modeling the control work in a similar way to a conventional synchronous generator [12]. The system works in parallel with a conventional base-load unit delivering 50 MW, with the BESS system taking care of the primary frequency control. The initial power of the BESS system is zero. The general information of the system and its controls is shown in Table 1.

Table 1. Concentrated model information.

Generating unit	Nominal power MW	Initial power MW	Inertia MWs^2/rad	droop
Conventional unit	50	50	3.16	No
BESS	50	0	0	0.1%

Two experiments are scheduled:

- Abrupt load increase of 30 MW.
- Variation of power generated by a photovoltaic solar farm working in parallel with the BESS system and the conventional unit having a generation pattern as shown in Fig. 4.

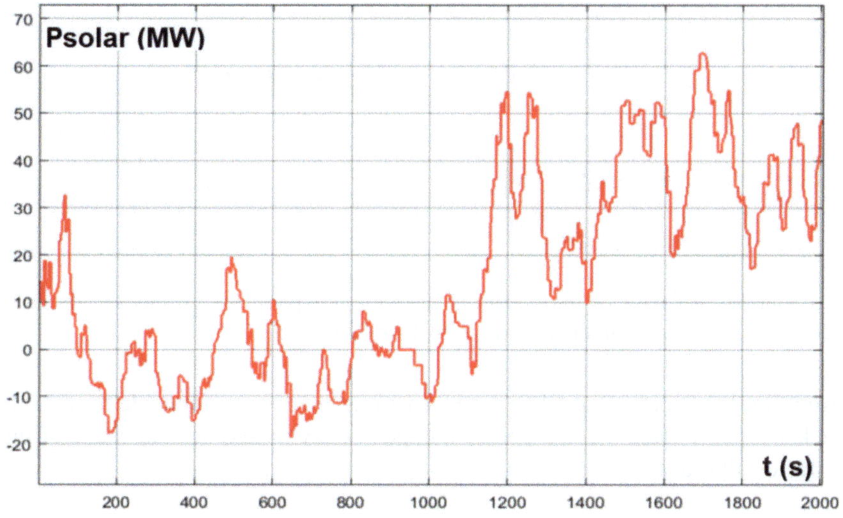

Fig. 4. Photovoltaic solar power

The results of the first experiment are shown in Fig. 5, which shows the frequency behavior of the system under the sudden disturbance of 30 MW load increase. In green is shown the response of the shaper inverters which practically coincide and in black the response of the follower inverter. The result is similar to that shown in reference [1]. The results of the second experiment for a BESS primary control deadband of 0.2 Hz and common droop in all cases of 0.1% are shown in Fig. 6.

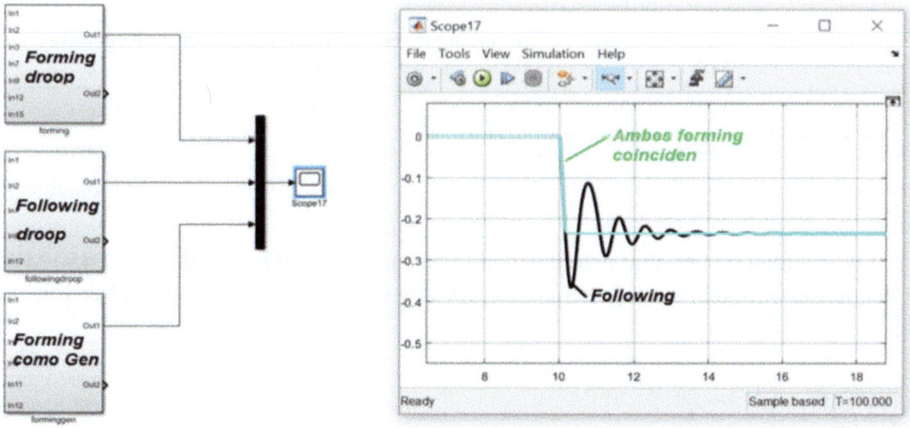

Fig. 5. Simulation results of the behavior of BESS systems with different types of inverters in the presence of a stepper type disturbance.

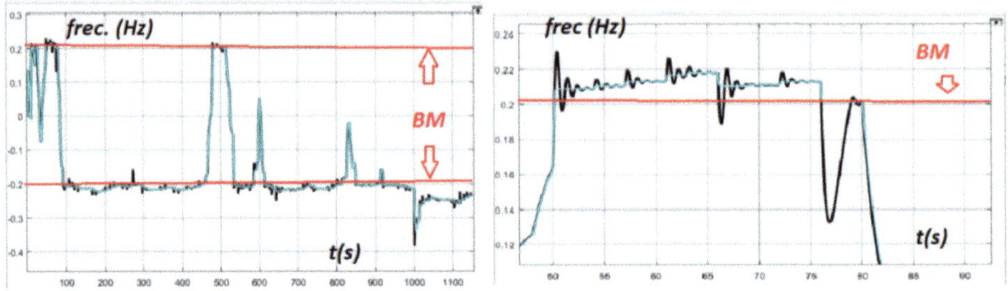

Fig. 6. Frequency behavior of the system in the second experiment, on the left, on the right enlargement of the behavior in the interval from 47 to 82 s

From Fig. 6 it is clear that when control actions are executed outside the dead band the tracker type inverters have a poorly damped behavior and are the cause of higher frequency oscillations.

5 Estimated Frequency Performance in December 2025 with the Addition of About 1200 MW of Solar PV Generation from New Farms Considering 80% Utilization

It is required to check the worst case scenario that may occur on that date, for which it is established that only the composition of generating units shown in Table 2 will be available.

Table 2. Units available to be used to cover the load for that date.

Generating unit	Nominal power MW	Minimum power MW	Maximum power MW	Regulating
Thermal	1137	623	853	MG6
Gas	317	242	317	Varadero
Hydro	15	2	14	Hanabanilla

It is assumed that by that date the load to be supplied will have a deficit of 450 MW in the period of higher solar generation according to the estimated forecast. See Fig. 7.

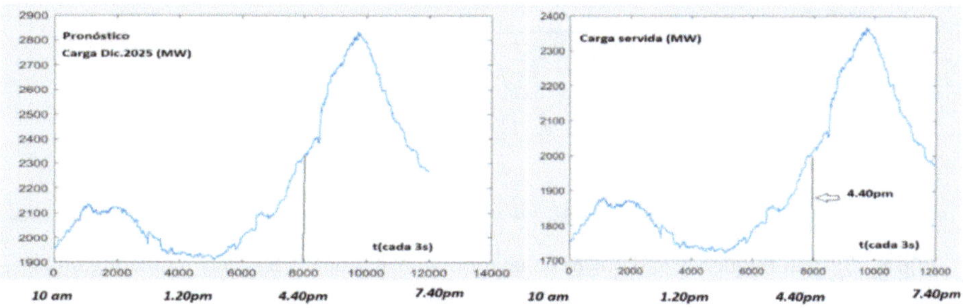

Fig. 7. On the left, forecast load to be served in December 2025 and load to be served as of 10 am.

Other considerations set for the study of frequency behavior:

- The availability of other types of generation is not considered. Load droop is considered as 2.44%. The inertia of the system according to the composition of units is 45 MWs2/rad.
- The maximum solar generation is considered as 80% of the total installed capacity, corresponding to 962 MW with the new parks and 116 MW with the old ones for a total of about 1100 MW.
- The solar parks have a reserve to regulate downward, when the frequency rises above 60.2 Hz, of 200 MW, with a droop of 1% and a delay of 0.5 s, a typical value for tracker type inverters. It is included that they regulate upwards also with 75 MW of reserve.
- The work of electric arc steel mills and wind farms is not considered.

  - There are no BESS systems for primary frequency control.
  - The Maximo Gomez 6 unit has a regulation reserve of ± 5 MW, Energas Varadero + 45 MW and the Hanabanilla Hydroelectric + 12 MW.
  - The frequency regulation droop of MG6 is 5%, of Energas Varadero is 2% and of Hidroelectrica is 7.5%.
  - Distributed generation units and contracted floating units are not considered in service.

These considerations are the worst that could occur at the date under analysis.

The results of the modeling performed under the above considerations are shown in Figs. 8 and 9.

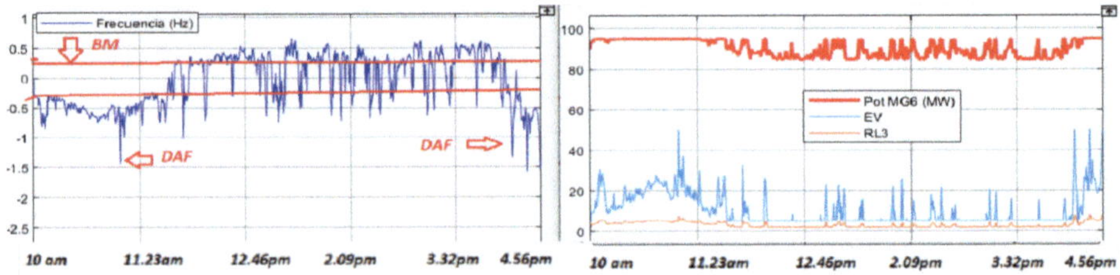

Fig. 8. System frequency and power behavior of conventional frequency regulating units.

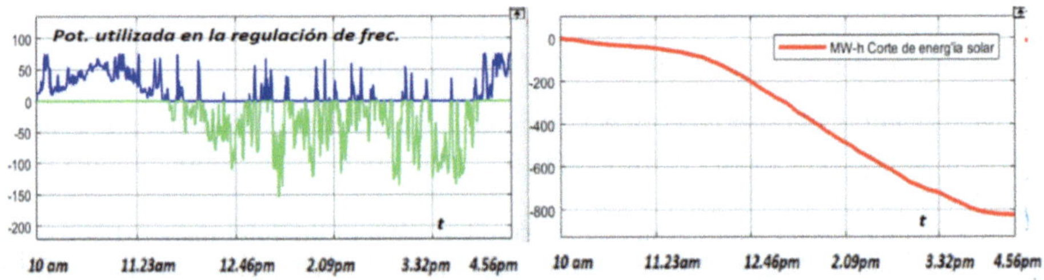

Fig. 9. Power moved by the wind farms in frequency regulation and unused energy.

The frequency has a behavior with frequent excursions out of the dead band, especially downwards in values between -0.5 and 1 Hz, which can cause DAF operation in the first delayed step before noon as shown in Fig. 9. In the case of the afternoon, the observed frequency drop can be solved by planned discharges, since the conventional generation capacity with the drop of the solar PV generation is not enough to cover the demanded load. In the case of the situation presented before noon, in order to avoid this discharge, it is necessary to increase the regulation reserve when the frequency drops. It is not convenient to do this by leaving more regulation reserve in the parks because this would be equivalent to significantly increase the unused solar energy, because for this it is necessary to work a set of parks below the maximum value indicated by the MPPT algorithm.

Figure 10 shows the solar power used in the system on the left and the load served on the right. As can be seen after 4 pm, it is necessary to continue discharging the system to avoid the frequency drop shown in Fig. 8.

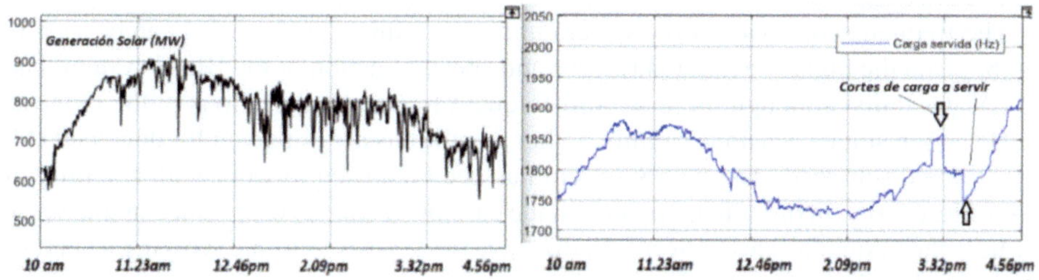

Fig. 10. Solar power used on the left, load supplied on the right with necessary outages due to generation deficit.

It is important to note that the result of the presented modeling implied the need to start load shedding as shown in Table 3.

Table 3. Load shedding

Time of load disconnection	Discharged power in MW
3.32 pm	50
4.07 pm	50

To avoid frequency deviations much higher than 60.2 it is necessary to offload two 50 MW steps of solar at 11.56 am and 12.56 pm which are reconnected at 3 and 3.15 pm.

6 Analysis of Frequency Behavior in Case of 100 MW, 100 MWh of BESS Systems for Primary Frequency Control

If the scenario analyzed above is analyzed, but now with the Energas Varadero combined cycle working in base regime and 100MW, 100MWh BESS systems for frequency control are incorporated to the frequency control, the ones planned to have a behavior that simulates a conventional generator, with 1% droop and working with 25 MWh setting, that is, 65 MWh to regulate when the frequency falls below the dead band (59. 8 Hz) and 15 MWh to regulate frequency above 60.2 Hz, the results obtained are shown in the following figures.

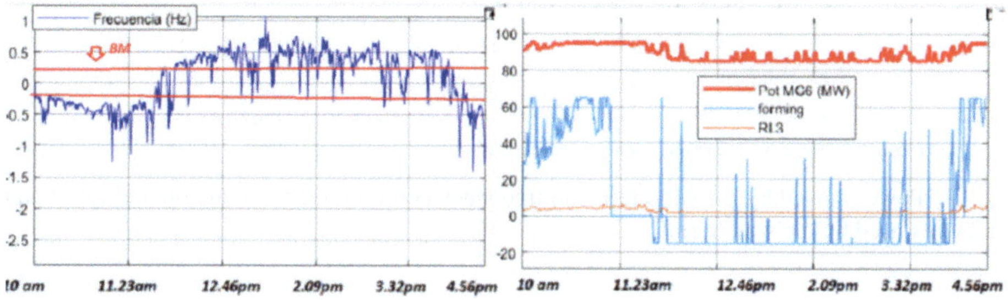

Fig. 11. Frequency and power performance of units under primary control, including BESS.

The frequency behavior of the system improves with respect to what can be seen in Fig. 8, given the higher regulation reserve and speed of action of the BESS system than the EV gas units, although there are still frequent frequency deviations in the range ± 0.5 Hz and in some occasions slightly higher, conditions are not given for delayed DAF steps to be discharged. The BESS system has been adjusted to work with a setting of 25% of its energy, in order to help more in the case of frequency drop, since in this sense is where the system has less reserve. In this case the EV gas units are working as part of a combined cycle and work in base regime.

In Fig. 11 on the right it can be seen that the BESS system reaches 10% of its energy and saturates, not taking part thereafter in frequency control, at about 11.23 am and joins the regulation at about 12m and is maintained until the end of the modeled time interval.

The BESS performs just under a full duty cycle as can be seen in Fig. 12. Here one can also see the time of park saturation and the time when it begins to regulate in the opposite direction as noted above.

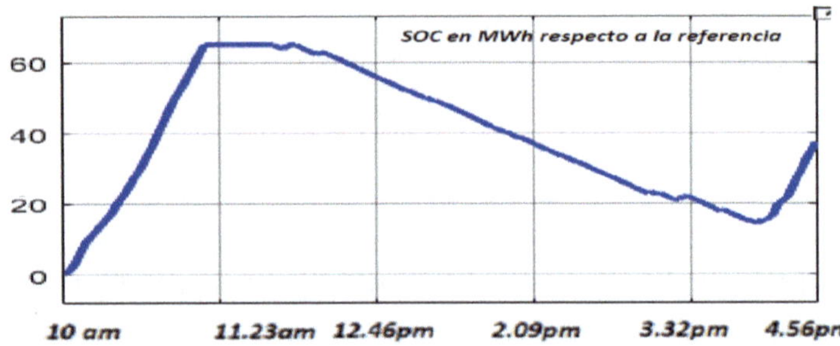

Fig. 12. SOC behavior of BESS systems. Reference 25% of the nominal energy.

The power used in the frequency regulation of the farms is shown in Fig. 13, as well as the energy that is not used due to the frequency regulation. In this case there is no need to disconnect the wind farms. The above leads to the fact that the regulation of the parks for frequencies above 60.2Hz behaves with control actions of higher magnitude.

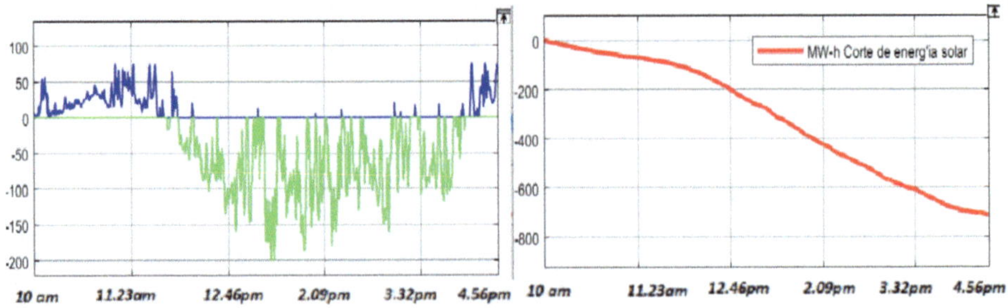

Fig. 13. Power moved by the wind farms in frequency control on the left and on the right Solar energy not used by the wind farms.

The energy not used by the farms is slightly lower than in the case where BESS systems are not used.

Figure 14 shows the power used by the solar photovoltaic generation, which has a similar behavior to the previous case on the left and the load served by the system on the right.

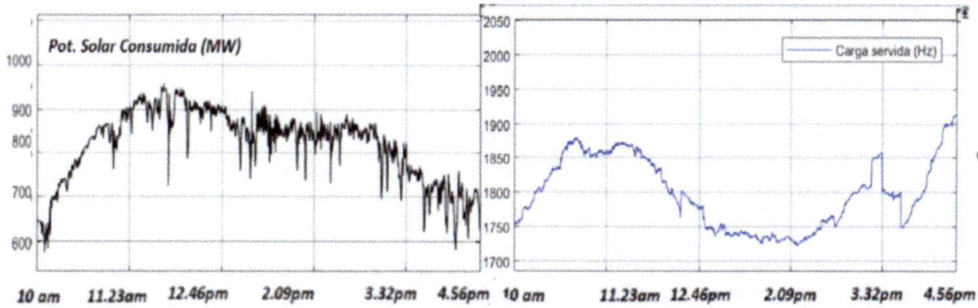

Fig. 14. Solar power used on the left, load supplied on the right with necessary outages due to generation deficit.

In this case there is no need to disconnect farms and the Energas Varadero cycle remains stable delivering 75MW all the time.

7 Conclusions

- The current situation of the Cuban SEP requires seeking primary regulation reserve to regulate frequency when it falls below 60-BM Hz. If this requirement is not fulfilled, the operating regimes of the system will be of low quality in terms of frequency behavior with undesired operations of some delayed steps of DAF, as well as oscillations that may endanger the stability of the system.
- The present work does not address the possible existence of oscillations between generating plants, for which it is necessary to carry out studies considering the configuration of the system network, with the objective that in advance of possible problems that may arise, the necessary measures are taken for their solution.

References

1. Lasseter, R., Chen, Z., Pattabiraman, D.: Grid-Forming Inverters: a Critical Asset for the Power Grid (2019)
2. Pereza, R., Hoffb, T., Diseb, J., Chalmersb, D., Kivalova, S.: The cost of mitigating short term output PV variability. 2013 ISES Solar World Congress. https://energypedia.info/images/7/75/Output_4._Guia_dimensionamiento_almacenamiento_GD_baterias.pdf
3. Puerto Rico Electric Power Authority (2012). Available: http://www.fpsadvisorygroup.com/rso_request_for_quals/PREP A_Appendix_E_PV_Minimum_Technical_Requirements.pdf
4. Lappalainen, K., Maki, A., Valealahti, S.: 28th European Photovoltaic Solar Energy Conference and Exhibition, pp. 4081 (2013)

5. Sanchez Reinoso, C.R., Milone, D.H., Buitgaro, R.H.: Appl. Energy **103**, 278 (2013)
6. Kern, E.C., Gulachenski, E.M., Kern, G.A.: IEEE Trans. Energy Convers. **4**, 184 (1989)
7. Lave, M., Kleissl, J., Arias-Castro, E.: High-frequency irradiance fluctuations and geographic smoothing. Sol. Energy **86**(8), 2190–2199 (2012)
8. Characterization of the spatio-temporal variations and ramp rates of solar radiation and PV. Report of IEA Task 14 Subtask 1.3 (2015). ISBN: 978-3-906042-35-0
9. Japan International Cooperation Agency (JICA): Proyecto para introducción de Energías Renovables en la Isla de la Juventud. Cuba (2019)
10. Martínez-García, A.A., Torres-Breffe, O.E., Vilaragut-Llanes, M., Delgado-Fernandez, O., Szpytko, J., Salgado Duarte, Y.: Solar Generation Estimation in Electric Power Systems for Prospective Frequency Control Studies. TransNav Int. J. Mar. Nav. Saf. Sea Transp. **16**(2), 211–217 (2022). https://doi.org/10.12716/1001.16.02.02
11. Guerra-Hernández, L., Martínez-García, A.A., Vilaragut-Llanes, M., Salgado-Duarte, Y., Szpytko, J.: Prospective Estimation of BESS Nominal Values for Frequency Control with High PV and Wind Integration. In: 2023 27th International Conference on Methods and Models in Automation and Robotics (MMAR), Międzyzdroje, Poland, pp. 314–319 (2023). https://doi.org/10.1109/MMAR58394.2023.10242553
12. Roos, P.: A Comparison of Grid-Forming and Grid-Following Control of VSCs. Universidad de Upsala, Dinamarca. Tesis de Doctorado (2020)

Design and Simulation of a Bidirectional Active Half-Bridge Dc-Dc Converter for V2G Applications

Raynel Díaz Santos(✉), Pablo Antonio Rodríguez Gutierrez, and Leonardo René Garcés Enriquez

Distributed Generation Research Group, Electrical Faculty of Technical University of Havana CUJAE, 10400 Havana, Cuba
raynelds@electrica.cujae.edu.cu

Abstract. The bidirectional active half-bridge converter (BAHB) is a versatile DC-DC topology enabling efficient energy transfer between sources like batteries, grids, and renewable systems. It employs dual half-bridge circuits coupled via a high-frequency transformer, providing galvanic isolation and bidirectional power flow control through phase-shift modulation and pulse-width modulation (PWM). Key advantages include reduced voltage stress on switches (handling half the input voltage), zero-voltage switching (ZVS) capabilities to minimize losses, and adaptive energy storage techniques to extend ZVS operability across load ranges. The paper presents the design and simulation of a bidirectional active half-bridge DC-DC converter for V2G applications. In the first stage, the topology is selected taking into account the minimum deviation of the output current, the size and weight of the output filter, the total output current harmonics deviations, and the minimum number of the component. In this sense, the modeling and simulation of the power converters is presented. Finally, the simulation results were obtained with the required performance in the operation of the power converter in both ways. The primary outcome of this research is the validation of the described design procedure, along with the refinement of design parameters to achieve higher efficiency.

Keywords: V2G applications · DC-DC converter · electric vehicles

1 Introduction

The elements that have enabled the boom in renewable energy sources (RES) are power electronic converters, which play a fundamental role in their integration into modern electrical systems. These devices act as a critical interface between variable generation sources (such as solar panels or wind turbines) and the technical requirements of the grid. They not only transform direct current (DC) into alternating current (AC)—as seen in photovoltaic inverters adapting solar panel output to grid parameters—but also optimize efficiency through techniques like maximum power point tracking (MPPT), maximizing energy extraction under varying irradiation or wind conditions [1, 2].

A. Santos Fuentefria et al. (Eds.): JIFREEEMA 2025, LNNS 1933, pp. 167–183, 2026.
https://doi.org/10.1007/978-3-032-23028-7_16

Furthermore, these converters ensure grid stability by managing harmonic distortions, imbalances, and voltage fluctuations—common challenges in systems with high RES penetration. Their ability to regulate active/reactive power ratios and provide ancillary services (e.g., frequency support or reactive compensation) makes them indispensable for a safe energy transition, reducing fossil fuel dependence and enabling decentralized management of distributed energy resources [1, 2].

In recent decades, power electronic converters have advanced significantly, serving as motor controllers in electric vehicles (EVs), and as battery chargers. Coupled with expanded charging infrastructure, this has driven global EV adoption, now integral to public and private transportation [3, 4].

Cuba has actively pursued this shift toward cleaner energy, addressing environmental goals and energy demands through nationally adapted technologies. Strategies leveraging charger bidirectionality and expanded charging points are projected for large-scale implementation, contributing to nearly one million EV sales in Cuba over five years [5]. Today, EVs are vital for the Cuban population, mitigating urban transport shortages while advancing climate goals and sustainable mobility.

Taking these principles into account, the main objective of this paper is to present the design of a bidirectional DC-DC converter with an active half-bridge topology, intended as a charger for lithium-ion batteries used in small electric vehicles such as motorcycles and tricycles. The paper outlines a complete design procedure for this type of converter.

The choice of power electronic converter topology for V2G applications is directly related to the initial design parameters, as it determines key aspects such as efficiency, bidirectionality capability, and grid compatibility. Factors like semiconductor selection, sizing of inductors and capacitors, and control strategies depend on the chosen architecture, ensuring that the system meets the stability, protection, and performance requirements demanded for effective vehicle-to-grid integration.

In this sense, the bidirectional half-bridge DC-DC converter shown in the figure combines boost and buck conversion functions within a single topology, allowing current to flow in both directions. This is achieved by using two semiconductor switches with antiparallel diodes that alternate conduction to reverse the direction of energy transfer.

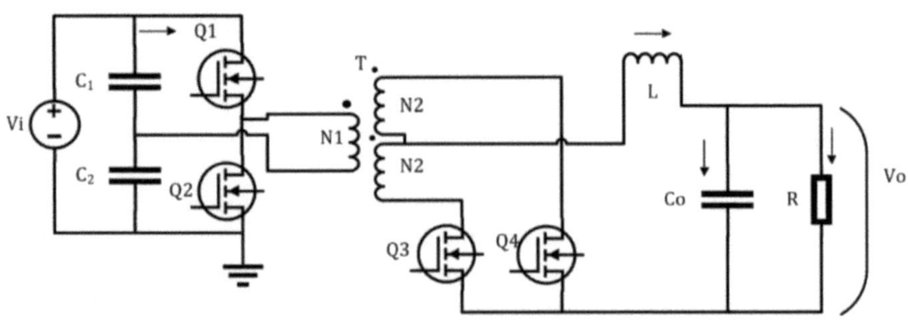

Fig. 1. Proposed bidirectional active half-bridge converter. **Source:** Own elaboration

When one switch is activated with a given duty cycle and the other is deactivated, the converter operates in boost mode, raising the output voltage; switching the active device and allowing current to flow through the opposite diode results in buck mode, where the voltage is reduced. This bidirectionality enables control of currents with both polarities in the converter's inductive coil [6].

In the next section we are talking about the design procedure and initial conditions and parameters to it needed for the following analysis.

2 Design Procedure and Initial Conditions

To carry out the design of the proposal converter, we can describe the procedure of the design step by step as follow:

1. Definition of initial conditions for design procedure
2. Topology selection
3. Steady State analysis for select all components of the converter
4. Validate the calculations of steady state by simulations

First, it is essential to define the initial design conditions, which include parameters such as the input and output voltage ranges, nominal power, switching frequency, and specific application requirements like bidirectionality and desired efficiency. This stage sets the foundation upon which the topology will be selected and the components sized, ensuring that the converter meets the functional and operational specifications [7–9].

Second, the appropriate converter topology is selected, taking into account factors such as the ability to operate in both buck and boost modes, circuit complexity, efficiency, and the capacity to handle bidirectional power flow. For example, half-bridge or full-bridge topologies are often preferred for V2G applications due to their flexibility and performance. This choice conditions the subsequent analysis and the selection of semiconductor devices, inductors, capacitors, and other passive components [7–9].

Finally, a steady-state analysis is carried out to select and size all the converter components, calculating key parameters such as inductance, capacitance, maximum currents, and expected losses. Subsequently, these theoretical calculations are validated through simulations that verify the dynamic behavior and efficiency of the system under real operating conditions. This validation is crucial to fine-tune the design before physical implementation, ensuring that the converter functions correctly and meets the established performance objectives [7–9].

Now, to begin the procedure and be able to perform the steady-state analysis of the converter, calculate and select the fundamental elements that comprise it, it is necessary to take into account another series of data and design criteria.

Table 1 shows the input and output parameters and designs criteria of the converter as we can see.

The table presents the key input parameters for the design and operation of a bidirectional DC-DC converter, specifying a DC bus voltage of 400 V, a switching frequency of 50 kHz, and ripple limits for both voltage (5%) and current (1%).

Table 1. Input and output parameters for design the converter. **Sources:** own elaboration

Parameter	Value	Parameter Value	Value
DC-Link	400 V	Power Output	1 kW
Switching Frequency SHBC	50 kHz	Output Voltage	84 V
Input Voltage deviation BAHB	5%	Output Current	12 A
Input Current deviation BAHB	1%	Efficiency	85%
Duty cycle	0.4	Input Voltage deviation BAHB	10%
		Input Current deviation BAHB	10%

These values reflect typical conditions to ensure efficient and stable operation of the converter, minimizing fluctuations that can affect power quality and system performance. The 400 V bus voltage is common in energy storage and electric vehicle applications, while the 50 kHz switching frequency represents a compromise between magnetic component size and switching losses. The established ripple levels ensure that voltage and current variations remain within acceptable ranges to protect components and maintain operating efficiency.

Additionally, the output power rating is 1 kW, with a maximum output voltage for the application in 84 V and a maximum current in 12 A. The initial design contemplates an efficiency of 85% and a duty cycle of 0.4, adjusted to operate with a nominal 72 V battery. These values reflect the operating conditions expected to ensure adequate performance, balancing power delivery, energy efficiency and compatibility with battery voltage, which is essential for applications such as energy storage systems or electric vehicles. With all those parameters and conditions, we can able to calculate other needed values like input power P_{in}, load resistance R_o, and also maximum and minimum input voltage ($U_{máx}$, U_{min}, respectively).

Therefore, we ready for second step, topology selection. Selecting the appropriate topology for bidirectional DC-DC converters is a fundamental aspect in the design of power conversion systems, especially in modern applications such as the integration of RES, battery storage, and electric vehicle management systems. Flyback, push-pull, half-bridge, and full-bridge topologies offer different advantages and limitations in terms of efficiency, cost, size, control complexity, and power handling capacity. The optimal choice depends on factors such as the required power range, space constraints, the need for galvanic isolation, and the system's efficiency and reliability objectives. A comparative analysis of these topologies allows the identification of the most suitable solution for each application, maximizing the technical and economic performance of the converter based on the specific demands of the operating environment [10–14].

After a detailed analysis of the main bidirectional converter topologies applicable to medium power chargers, where classic topologies such as flyback, push-pull, full bridge and half-bridge converters were studied, evaluating their technical characteristics, efficiency, control complexity, size, cost and galvanic isolation capacity [10–14].

With special emphasis on 1 kW systems oriented to applications such as electric vehicles and energy storage, it was shown that topologies such as flyback and push-pull present significant limitations for the target power, due to high losses, complexity in transformer design and lower operational efficiency, making them not very recommendable for a 1 kW bidirectional charger. On the other hand, although the full bridge offers high robustness and efficiency, its higher number of switches and the full voltage applied to the transformer increase system complexity, cost and switching losses, aspects that can be critical in medium power applications. However, the bidirectional half-bridge converter stood out as the most suitable topology for this application due to its balance between efficiency, simplicity, and cost. This topology uses only two switches, which significantly reduces switching losses and simplifies control. Furthermore, the transformer operates at approximately half input voltage, allowing for a more compact and economical design of the magnetic core [15–20].

Additionally, half-bridge offers efficient management of bidirectional power transfer, key aspects for safety and performance in battery chargers. Therefore, the selection of this topology represents an informed decision that optimizes the balance between performance, cost and complexity, forming the basis for the successful development and implementation of the proposed system [15–20].

So, from the two previous aspects of the design procedure, the steady-state analysis of the converter can be performed, taking into account the operating principle of the topology selected [15–20].

2.1 Operating Principle

Figure 1 shows the circuit diagram of the proposed BAHB converter. The design uses a half-bridge connected with the DC power supply on the primary side and a center-tapped transformer and a synchronous rectification on the secondary side [10]. The converter can operate in two modes, step-down mode and step-up mode and all of the four switches Q1–Q4 are gated in both modes [10]. Switch Q1 with switch Q4, and also switch Q2 switch Q3 are complementary.

In the step-down mode, the primary side DC power supply Vi (400 V) charges the secondary side battery Vo (84 V), and Q3 and Q4 provide rectification. In contrast, Q1 and Q2 operate as rectifiers in the step-up mode when Vo supplies the high side battery Vi [10]. Some assumptions are made for mathematical insight into the proposed converter, as: (i) the ON-state resistance of all switches is ignored; (ii) the capacitors C1, C2 and Co are large enough [10], and the voltage across the capacitors can be taken as constant; and (iii) the capacitance of the capacitors C1 and C2 is equal i.e. C1 = C2 = C [10]. Thus, VC1 = VC2 = Vi/2; and (iv) The diode forward resistance is zero in the body diodes of the transistors Q1–Q4 [10]. The behavior principle of the BAHB was developed in [10], describing the operation stages of step-down and step-up modes.

With the operating principle describe before and also the initial conditions the steady state analysis was made. The results of the analysis will present in the next subsection.

2.2 Steady State Analysis

In this section, we aim to calculate all the necessary values of steady state of the converter, which in turn allows us to determine the value of each of the essential elements of it.

In [21], the volts-second balance in half-bridge converter can be expressed as follows:

$$\left(\frac{U_{in}}{2 \cdot n} - U_o\right) \cdot D = U_o \cdot \left(\frac{1}{2} - D\right) \cdot T \tag{1}$$

where:

U_{in} → Input Voltage
U_o → Output Voltage
n → Transformer turns ratio
D → Duty Cycle
T → Switching period (1/fs)
And the voltage transfer function (M_{VDC}) can be written as follow if the D < 0.5

$$M_{VDC} \equiv \frac{U_o}{U_{in}} = \frac{I_{in}}{I_o} = \frac{D}{n} para\ D \leq 0.5 \tag{2}$$

With the values of table 1, Eq. 2 can be evaluated, and with these results we can calculate the maximum turns ratio of the converter, where is equal to 1.923. The choice of the parameter is n = 1.5. Furthermore, with M_{VDC} and n, the calculation of duty cycle is equal to [21, 22]:

$$U_o = \frac{U_{in} \cdot D}{n} \tag{3}$$

With this equation, we can obtain the nominal duty cycle, that it is equal to 0.319 for maximum load. Also, it is possible to obtain maximum and minimun values of D, 0.336 and 0.304 respectively. With all of those, we can calculate the input capacitance, the output capacitance, the minimum inductance for continuous conduction mode (CCM), the parameter of the transformer with selection of the core and the parameter to select the power semiconductors as follow.

Selection of Input Capacitance (Ci)

When selecting the input capacitor, the voltage and current to which it will be exposed must be taken into account. The input current is obtained by the ratio between the input power and the minimum input voltage, being 3.158 A. The voltage ripple to which it will be subjected was defined within the design criteria, 5% in Table 1, and is equal to 40 V, since it ranges from a minimum of 380 V to a maximum of 420 V [15, 21].

$$Ci_{min} = \frac{I_{in}}{2 \cdot F_{line} \cdot \Delta U} = 657.895\ \mu F \tag{4}$$

where Iin is the input current, ΔU is the voltage ripple and F_{line} the frequency undulation of the DC-link (normally 60 Hz for our electric power systems). Also, the maximum voltage across the capacitors is Vin maximum divided by 2, in this case 210 V. Taking into account the results, and knowing that the ESR for the input capacitance is 1 Ω, the selected capacitors are 820 μF and 250 V from EPCOS capacitors datasheet.

Selection of Output Capacitance Co

In the case of the output capacitor, a similar procedure to that of the input capacitor can be followed, which in this case the frequency that will be affecting this capacitor is the switching frequency (fs). In addition, the output voltage ripple was selected 1% of the output voltage in Table 1. Then, the minimun output capacitance Co can be calculated as indicate [15, 21].

$$Co_{min} = \frac{I_o}{\Delta U_o \cdot 2 \cdot fs} = 142.857\ \mu F \tag{5}$$

where Io is the output current (12 A), ΔUo is the voltage ripple (0.84 V), and fs the switching frequency (50 kHz). Also, the maximum voltage across the capacitors is Vo, in this case 84 V. Taking into account the results, and knowing that the ESR for the output capacitance is 0.110 Ω, the selected capacitor is 200 μF and 100 V from EPCOS capacitors datasheet.

Calculation of Minimum Inductance Lo for CCM

The output inductance defines the operating mode of the converter, distinguishing between CCM and discontinuous conduction mode DCM, with the critical inductance being crucial to calculate to avoid unwanted transitions that compromise the stability and efficiency of the system.

To design an inductor, the procedure was in [15, 21, 23] and the abstract of it can be describe as follow. The minimum inductance value (Lomin) is first calculated, which defines the boundary between CCM and DCM, applying Faraday's law to ensure the inductor operates correctly in the desired mode. Next, the Area product (Ap) is determined, which is essential for selecting the appropriate core that meets the required electrical and thermal characteristics. Based on Ap, the inductor core is chosen, providing key values such as permeability (Λ) and the A_L constant (in nH per turns squared), necessary for precise winding dimensioning [15, 21, 23].

Then, the number of turns of the inductor is calculated using the A_L constant and the required inductance value, ensuring the design meets the electrical specifications. The conductor gauge for the winding is also determined, considering the maximum current it must carry, and the equivalent series resistance of the inductor is calculated, a factor that affects losses and converter efficiency [15, 21, 23].

To validate the design, the actual available core area ($K_w A_w$) is compared with the area occupied by the winding, ensuring sufficient space for the coil without compromising mechanical or thermal integrity. Finally, the actual current ripple is calculated based on the selected inductance (Lo), allowing evaluation of the inductor's dynamic behavior during converter operation [15, 21, 23]. A briefly abstract of the process can be seen in Table 2.

Where Lo_{min} is the minimum inductance at the boundary between CCM and DCM, and Ap is the area product—used first to select the appropriate core and then to ensure that the core meets the requirements, N_L is the calculated and selected number of turns,

Table 2. Inductor parameters. **Sources:** own elaboration

Parameter	Calculated Value	Selected Value
Lo_{min}	0.321 mH	0.5 mH[1]
Ap	21 cm^4	[2]25.911 cm^4
N_L	56.796	57
A_{st}	2.668 mm^2	[3]3.308 mm^2
L_{wire}	8.312 m	8.312 m
R_L	0.044 Ω	0.044 Ω

A_{st} is the cross-section area of the wire used for winding, L_{wire} is the length of the winding and R_L is the electrical resistance of the winding.

To check the inductor design based on its geometric capabilities, the following expression is used:

$$K_w \cdot A_w > A_{st\_s} \cdot N_L \tag{6}$$

where K_w is the window factor (0.6 for inductors), A_w is the window area, A_{st_s} is the cross-section area of the wire used for winding and N_L is the number off turns.

$$K_w \cdot A_w = 6.003\ cm^2 > A_{st_s} \cdot N_o = 1.886\ cm^2$$

$$6.003\ cm^2 > 1.886\ cm^2 \therefore \textit{The inequality is satisfied}$$

Then, the inductor design meets all the requirements that were foreseen. After that, we can calculate the real current ripple in the output.

$$\Delta iL_o = \frac{U_o \cdot (0.5 - D_{min})}{f_s \cdot L_o} = 0.494\ A \tag{7}$$

where ΔiLo is the new output current ripple calculated by the inductor design.

The Parameter of the Transformer with Selection of the Core

The design of high frequency transformer has the same steps of the inductor design. Follow the previous analysis we can obtain the parameters of the design. In Table 3 we can see the results.

Where Ap is the area product—used first to select the appropriate core and then to ensure that the core meets the requirements (PQ 50/50 TDK EPCOS), N_p, N_p is the calculated and selected number of turns primary and secondary windings, A_{stp}, A_{sts} is the cross-section area of the wire used for each winding respectively, R_p, R_s is the

[1] The value of inductance that ensure CCM.
[2] Area product calculated by the datasheet parameters or core T225-52B.
[3] Cross section area of AWG 12.

Table 3. Transformer parameters. **Sources:** own elaboration

Parameter	Calculated Value	Selected Value
Ap	2.5 cm^4	11.289 cm^4
Np	24.439	24
Ns	16	16
A_{stp}	0.589 mm^2	0.6503 mm^2
A_{sts}	0.667 mm^2	0.820 mm^2
R_p	0.126 Ω	0.126 Ω
Rs	0.067 Ω	0.067 Ω
L_{Mmin}	1.857 mH	50 mH

electrical resistance of each winding and L_{Mmin} is magnetizing inductance calculated and selected.

To check the transformer design based on its geometric capabilities, the following expression is used:

$$K_W \cdot A_W > A_{st_p} \cdot N_p + A_{st_s} \cdot N_s \tag{8}$$

where K_w is the window factor (0.4 for transformers), A_w is the window area, A_{stp}, A_{sts} is the cross-section area of the wire used for each winding and N_p, N_s is the number off turns for primary and secondary windings.

$$K_W \cdot A_W = 1.36\ cm^2 > A_{st_p} \cdot N_p + A_{st_s} \cdot N_s = 0.288\ cm^2$$

$$1.36\ cm^2 > 0.288\ cm^2\ \therefore\ \textit{The inequality is satisfied}$$

Select the Power Semiconductors

In a power electronics converter, the main parameters for selecting power semiconductor elements are the maximum reverse peak voltage and the maximum current that will flow through the device. The maximum reverse peak voltage determines the semiconductor's ability to withstand transients and shutdown conditions without risk of damage, while the maximum circulating current ensures that the device can handle load peaks without overheating or compromising its integrity. In this sense, and following the indications of [15, 21] the expressions to determine these parameters are:

$$Us1_{max} = U_{max} = 420\ V \tag{9}$$

$$Is1_{max} = \frac{I_M}{n} = 8.4\ A \tag{10}$$

where $Us1_{max}$ is the maximum voltage across the transistor Q1 and $Is1_{max}$ is the maximum current through the Q1. The same analysis will be made for the power semiconductors in the secondary side where the MOSFET 20N60 from the manufacturer

Unisonic Technologies CO., LTD, with TO-247 encapsulation, is selected, which has as fundamental characteristics as a maximum drain source voltage Vdss = 600 V and maximum drain current Id = 20 A, as well as a conduction resistance (Rds(on) = 0.32 Ω). With these calculations, the majority of the component selection for this converter is complete, allowing us to proceed with calculating the power loss in each selected component.

Power Loss of Each Component in the Converter

These losses are obtained through detailed analysis of the converter's operating cycles, applying mathematical models that consider the electrical characteristics of the devices and the operating conditions, or through advanced electrical simulations that allow estimating both the instantaneous and average losses in steady state [15, 21]. The total system losses are simply the losses associated with each of the elements selected in this study. Mathematically speaking, this would be as follows:

$$P_T = \sum_{i=1}^{k} P_k \tag{11}$$

where P_T is the total losses of converter, and P_k is the losses related of each component. For this particular case we can rewrite the before equation in this sense:

$$P_T = P_{T\_rds\_sw} + P_{Ttrafo} + P_{Tinduc} + 2 \cdot PrCi + PrCo = 128.574\ W \tag{12}$$

where P_T is the total losses of converter, P_{Trdsw} is the losses related with the conduction and switching losses for four transistors (63.048 W), P_{Ttrafo} total losses in the transformer, P_{Tinduc} total losses in the inductor, PrCi losses related with ESR in input capacitance, PrCo losses related with ESR in output capacitance.

The efficiency of any system is calculated as the output power divided by the input power, where the input power is simply the output power plus the losses if analyzed per unit, but it is generally expressed as a percentage, therefore this division must be multiplied by 100. Expressed mathematically, it is:

$$\eta(\%) = \frac{Po}{Po + P_T} \cdot 100 = 88.806\% \tag{13}$$

where Po is the output power and P_T is the total losses. With an efficiency of 88.806%, it is evident that, although the converter presents a good level of performance, there are significant losses that must be identified and minimized to improve system operation and reliability.

3 Simulation Results Analysis

The close match between the simulated and theoretical waveforms validates the mathematical models used for steady-state operation, where variables are assumed to remain constant throughout each switching period, disregarding initial transients. This theoretical foundation is essential to ensure that the analyses and predictions of the converter's behavior are reliable and accurately represent its actual performance under stable conditions.

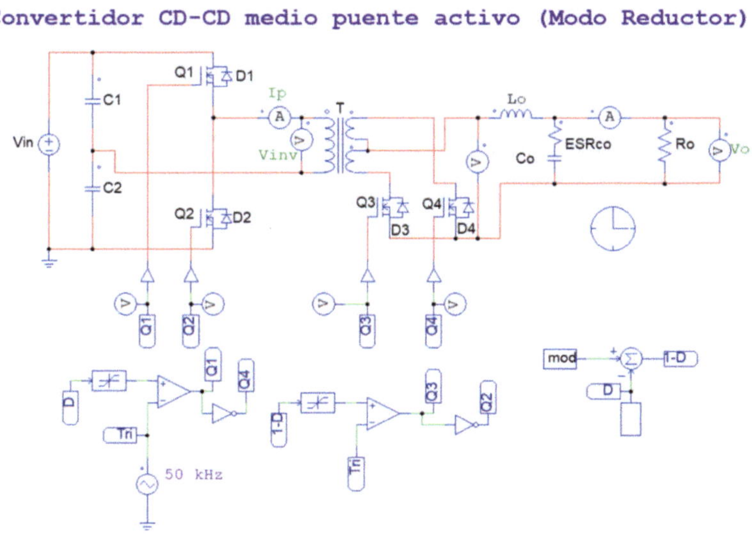

Fig. 2. BAHB in buck mode **Source:** Own elaboration

Figure 2 shows the electrical circuit of the proposal converter with PSIM software. To operate the converter correctly, the PWM modulation shown in the figure is required, which has a switching frequency of 50 kHz, and where a normalized triangular carrier signal (oscillating between 0 and 1, with a 50% duty cycle) is compared with two modulators of D and 1-D value. The result of the comparison are the activation signals of each of the power semiconductors that it can see in Fig. 3.

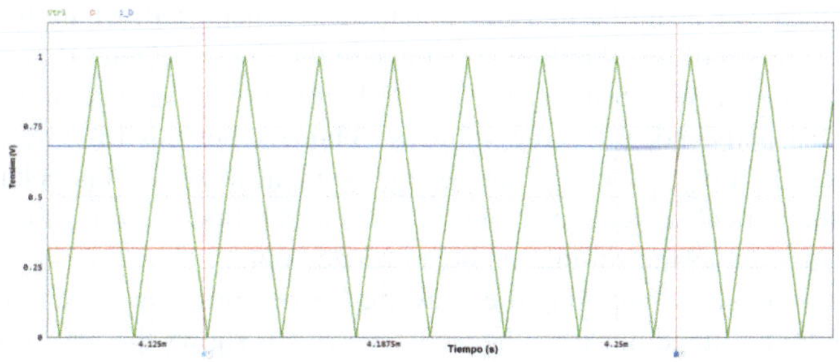

Fig. 3. PWM signal comparison. **Source:** Own elaboration

Now, the result of this comparison is shown in the figure below, taking into account the 4 semiconductors.

Figure 4 shows the activation signals for each of the converter's semiconductors. This achieves the required dead time between the conduction output of transistor Q1 and the conduction input of transistor Q2, ensuring proper device operation and matching the theoretical waveforms described in [21, 22].

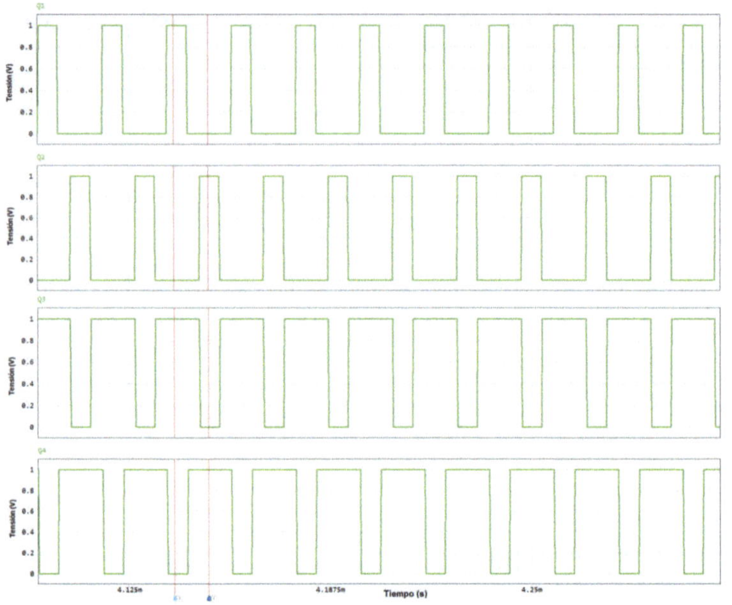

Fig. 4. PWM signal of each power semiconductor. **Source:** Own elaboration

Furthermore, it can be seen that Q1 and Q4 are complementary, as are Q2 and Q3; that is, when one is activated, the other is deactivated, and vice versa. The simulation for buck mode is analyzed in detail, but the activations achieved for the semiconductors are also valid for boost mode.

By activating and deactivating the semiconductors Q1 and Q2 in the transformer primary, a voltage wave similar to a sinusoidal wave is achieved, which is necessary to achieve the excitation of this winding, as shown in the Fig. 5. When Q1 is activated, the voltage at the transformer's primary is equal to half the input voltage in positive mode, and when Q2 is activated, the same value is obtained, but this time negative. When neither Q1 nor Q2 are conducting, the voltage is equal to zero. This behavior exactly matches the theoretical waveform shown in [21], which first validates the modeling of the converter, as well as the simulation being carried out.

In this same sense of validating the converter's operation, the current as well as the converter's output voltage must be checked. The Fig. 6 shows the waveforms of both. That figure shows the simulated waveforms of output voltage and current (Vo, Io) versus their reference values (Vref, Iref) in the converter, demonstrating the dynamic response of the system to an initial step.

An overshoot and damped oscillation can be observed in both voltage and current, followed by stabilization around the reference values, as a second-order system behaves.

The measurement table highlights that, after the transient mode, the simulated average output voltage and current values practically match the reference values, thus validating the model's accuracy and the correct tuning of the system under steady-state conditions. It is worth noting that the boost mode behaviors are similar.

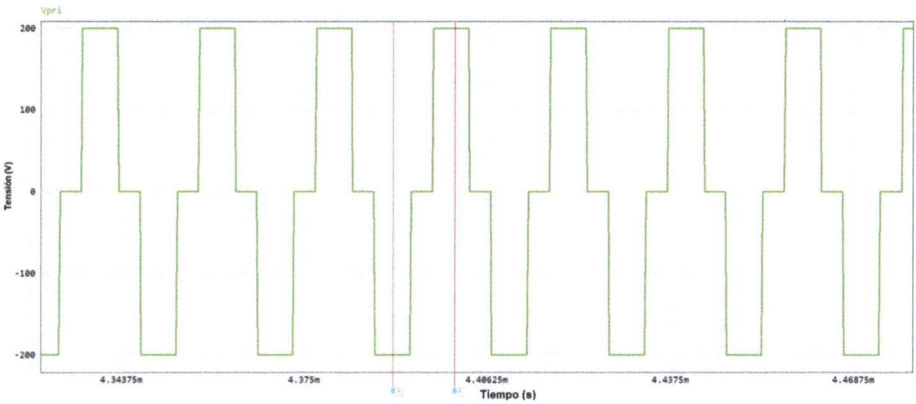

Fig. 5. Voltage at primary winding in the transformer. **Source:** Own elaboration

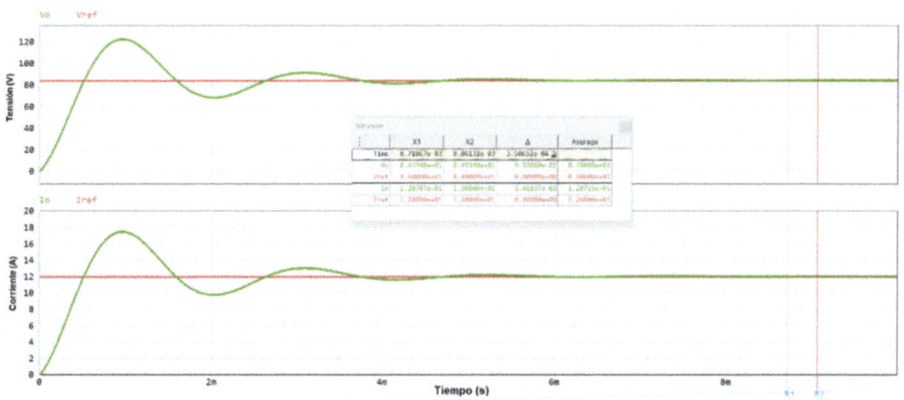

Fig. 6. Output current and output voltage. Source: Own elaboration

We can check if the converter is CCM mode, and also the current ripple with the inductor current waveform. That it is possible with Fig. 7.

Looking at the Fig. 7, we can see that the converter operates in CCM, because the current through the inductor never reaches zero. Furthermore, the maximum ripple from the simulation is 0.6 A, slightly higher than the calculated value (0.494 A). This may be because, even at this stage of the simulation, all the converter's parasitic resistances are being ignored, so the result is considered adequate.

Another interesting result to highlight is how the current is formed by the inductor at the output from the switching of Q3 and Q4, and the voltage that both semiconductors must block when they enter the off state. As observed in Fig. 8, in each switching of the transistors on the secondary side, that is, Q3 or Q4, the current in the output inductor is circulating through one of them respectively.

When Q3 and Q4 coincide, the currents are equal and decrease until one of the two leaves conduction. With respect to the voltage seen in transistor Q3 (VQ3) for example, it can be seen that, when the transistor is conductive its voltage drop is zero, and when

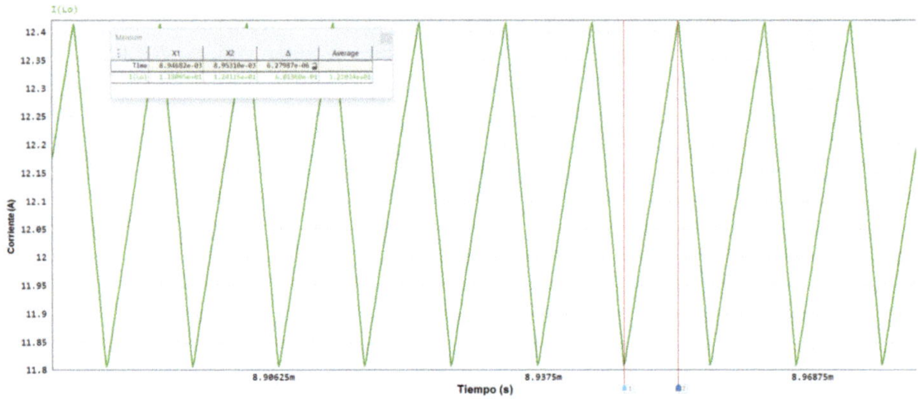

Fig. 7. Inductor current. Source: Own elaboration

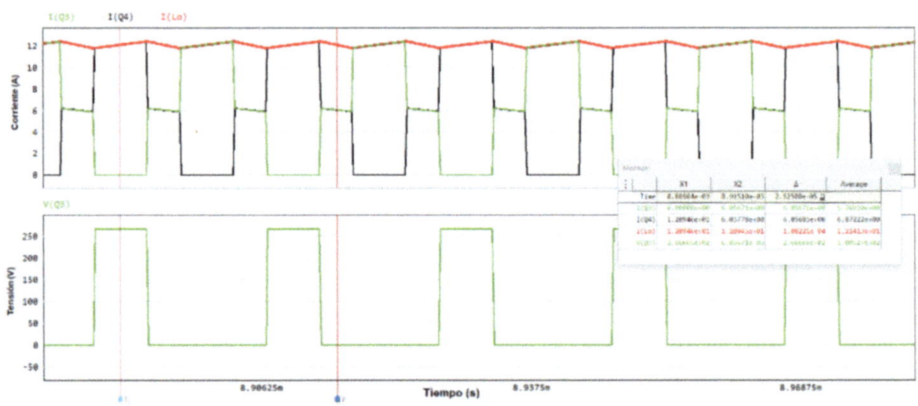

Fig. 8. Inductor current. Source: Own elaboration

it is off it has to withstand a voltage of 267 V, which corresponds to the primary voltage divided by the transformation ratio in this case.

Finally, a dynamic analysis can be performed in which the following is done:

1. A disturbance in the input voltage Vin that brings it to its maximum value of 420 V at 6 ms of simulation.
2. Verify how the output voltage and current values ncrease.
3. After 12 ms, decrease the duty cycle to its minimum value (Dmin = 0.304).
4. Verify how the output voltage and current values return to their initial condition.

The results of the dynamic analysis performed on the active half-bridge converter are presented, demonstrating the system's ability to respond effectively to input voltage disturbances and duty cycle adjustments. When the input voltage is increased to 420 V, an immediate increase in the output voltage and current is observed, reflecting the direct influence of the input on the output under transient conditions. Subsequently, when the duty cycle is reduced to its minimum value (Dmin = 0.304), the output variables return to their initial values, demonstrating the effectiveness of PWM control in maintaining the regulation and stability of the converter against external variations.

These results confirm the robustness and stability of the active half-bridge converter, showing fast response times and no sustained oscillations during transients. The ability to recover nominal conditions after disturbances ensures reliable and efficient operation, key aspects for applications requiring high precision and stability in power conversion. In this way, the dynamic analysis validates the system design, ensuring that the variables involved in the mathematical model used in the implementation of the control strategy have their correct values in steady state. These are fundamental aspects for optimizing the performance and durability of the converter under real-world operating conditions once its control loops are implemented (Fig. 9).

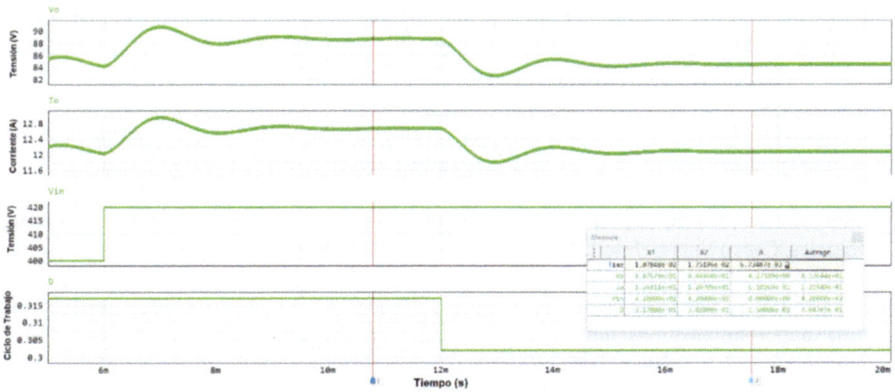

Fig. 9. Inductor current. Source: Own elaboration

4 Conclusions

A comprehensive and systematic study of bidirectional DC-DC converter topologies was conducted. A selection was made based on a comparison of the advantages and disadvantages of each topology for the specific application of this case study. A theoretical design of the selected topology was then carried out, covering all its components, and the results were validated by comparing theoretical waveforms with those obtained from simulation. This validation confirms that the case study is prepared to proceed with the design process described, ensuring that all steady-state variables have accurate values for modeling and control loop development. Therefore, it can be concluded that a bidirectional active DC-DC half-bridge converter has been successfully designed.

References

1. Blaabjerg, F., Yang, Y., Kim, K.A., Rodriguez, J.: Power electronics technology for large-scale renewable energy generation. Proc. IEEE **111**, 335–355 (2023). https://doi.org/10.1109/JPROC.2023.3253165

2. Tang, Z., Yang, Y., Blaabjerg, F.: Power electronics: the enabling technology for renewable energy integration. CSEE J. Power Energy Syst. **8**, 39–52 (2022). https://doi.org/10.17775/CSEEJPES.2021.02850
3. Kotb, R., Chakraborty, S., Tran, D.-D., Abramushkina, E., El Baghdadi, M., Hegazy, O.: Power electronics converters for electric vehicle auxiliaries: state of the art and future trends. Energies **16**, 1753 (2023). https://doi.org/10.3390/en16041753
4. Islam, R., Rafin, S.M.S.H., Mohammed, O.A.: Comprehensive review of power electronic converters in electric vehicle applications. Forecasting **5**, 22–80 (2023). https://doi.org/10.3390/forecast5010002
5. Valdés, A.L., Taño, M.A., Llanes, M.V., Santos, R.D., Vera, D.J.: The electric bicycle in Cuba. Recommendations for its use. Ing. Energética **43**, e3001 (2022)
6. Flores Oropeza, L.A.: Estudio y análisis de soluciones tipológicas de convertidores cc-cc bidireccionales para su aplicación en vehículos híbridos (2004). https://dialnet.unirioja.es/servlet/tesis?codigo=2620
7. Ekici, E., Koroglu, T., Çelik, Ö.: Design and analysis of two-stage bidirectional power converter for vehicle-to-grid technology with fuel cell-battery electric vehicle. J. Energy Storage. **106**, 114792 (2025). https://doi.org/10.1016/j.est.2024.114792
8. Martínez-Vera, E., Bañuelos-Sánchez, P., Rosado-Muñoz, A.: Design and validation of a bidirectional DC-DC converter control for electric vehicles using FPGA-in-the-loop methodology. J. Eur. Systèmes Autom. **57**, 77–85 (2024). https://doi.org/10.18280/jesa.570108
9. Wang, J., Wang, B., Zhang, L., Wang, J., Shchurov, N.I., Malozyomov, B.V.: Review of bidirectional DC–DC converter topologies for hybrid energy storage system of new energy vehicles. Green Energy Intell. Transp. **1**, 100010 (2022). https://doi.org/10.1016/j.geits.2022.100010
10. Adegbohun, F., von Jouanne, A., Agamloh, E., Yokochi, A.: A review of bidirectional charging grid support applications and battery degradation considerations. Energies **17**, 1320 (2024). https://doi.org/10.3390/en17061320
11. Gontijo, G., Wang, S., Kerekes, T., Teodorescu, R.: Performance analysis of modular multilevel converter and modular multilevel series converter under variable-frequency operation regarding submodule-capacitor voltage ripple. Energies **14**, 776 (2021). https://doi.org/10.3390/en14030776
12. Panchanathan, S., et al.: A comprehensive review of the bidirectional converter topologies for the vehicle-to-grid system. Energies **16**, 2503 (2023). https://doi.org/10.3390/en16052503
13. Kempton, W., Tomić, J.: Vehicle-to-grid power implementation: from stabilizing the grid to supporting large-scale renewable energy. J. Power. Sources **144**, 280–294 (2005). https://doi.org/10.1016/j.jpowsour.2004.12.022
14. Suresh, K., Arulmozhiyal, D.R.: Design and implementation of Bi-directional DC-DC converter for wind energy system. Circuits Syst. **7**, 3705–3722 (2016). https://doi.org/10.4236/cs.2016.711311
15. Pressman, A.: Switching Power Supply Design. McGraw-Hill Education (2009)
16. Rosso, R., Wang, X., Liserre, M., Lu, X., Engelken, S.: Grid-forming converters: control approaches, grid-synchronization, and future trends—a review. IEEE Open J. Ind. Appl. **2**, 93–109 (2021). https://doi.org/10.1109/OJIA.2021.3074028
17. Zhu, T., Zhao, F., Wang, X., Torrico-Bascope, G.V.: Adaptive harmonic conductance control for boost pfc converters at light loads. IEEE Trans. Power Electron. **39**, 3175–3185 (2024). https://doi.org/10.1109/TPEL.2023.3345936
18. Wang, X., Blaabjerg, F.: Harmonic stability in power electronic-based power systems: concept, modeling, and analysis. IEEE Trans. Smart Grid. **10**, 2858–2870 (2019). https://doi.org/10.1109/TSG.2018.2812712
19. Díaz Santos, R.: Estudio de un convertidor bidireccional para la conexión de los vehículos eléctricos a la red (2013). http://tesis.cujae.edu.cu:8080/handle/123456789/5920

20. Life-cycle energy demand comparison of medium voltage Silicon IGBT and Silicon Carbide MOSFET power semiconductor modules in railway traction applications. Power Electron. Devices Compon. **6**, 100050 (2023). https://doi.org/10.1016/j.pedc.2023.100050
21. Kazimierczuk, M.K.: Pulse-Width Modulated DC-DC Power Converters. John Wiley & Sons Ltd, The Atrium, Southern Gate, Chichester, West Sussex, PO19 8SQ, United Kingdom, Wright State University, Dayton, Ohio, USA (2016)
22. Arsalan, A., Puyang, C., Kim, H.-J.: A 3 kW bidirectional DC-DC converter for electric vehicles. J. Electr. Eng. Technol. **11**, 860–868 (2016). https://doi.org/10.5370/JEET.2016.11.4.860
23. Umanand, Z., Bhat, S.P.: Design Of Magnetic Components For Switched Mode Power Converters. New Age International (P) Limited (1992)

Lactic Acid Production from Hydrothermally Pretreated Sugarcane Bagasse Using *Pediococcus Pentosaceus* and a Thermally Treated Anaerobic Sludge

Beatriz Guerra-Alvarez[1], Chantelle Romero-Barrios[1], Köhler Stefan[3], Ileana Pereda-Reyes[2], and Ania Cabrera-Díaz[2](✉)

[1] Center of Studies the Process Engineering, Chemical Engineering Faculty, Technological University of Havana "José Antonio Echeverría (CUJAE), Calle 114 No. 11901, Marianao, La Habana, Cuba
beatrizg@quimica.cujae.edu.cu

[2] Study Center of Renewable Energy Technologies (CETER), Mechanical Engineering Faculty, Technological University of Havana "José Antonio Echeverría (CUJAE), Calle 114 No. 11901, Marianao, La Habana, Cuba
{ileana,ania}@mecanica.cujae.edu.cu

[3] Institute for Agricultural and Urban Ecological Projects (IASP), Humboldt University of Berlin, Berlin, Germany
stefan.d.koehler@iasp.hu-berlin.de

Abstract. Sugarcane bagasse (SCB) is one of the main residues of the sugar agroindustry; its complex lignocellulosic structure limits its biodegradation and utilization in the production of bio-based chemicals. This study evaluated the effect of hydrothermal pretreatment of raw SCB (HPSCB; 140 °C, 15 min; 62.3% delignification) on lactic acid (LA) production. Enzymatic hydrolysis of HPSCB using *Aspergillus niger* enzymes achieved a maximum reducing sugar concentration of 11.98 g/L under optimal conditions (50 °C, 80 PFU/g-SCB_{DW}, 24 h). Subsequently, LA production was investigated via pre-hydrolysis followed by simultaneous saccharification and fermentation (SSF) using two microbial sources: the pure strain *Pediococcus pentosaceus* LB-25 and thermally treated anaerobic sludge (TTAS). Key variables included microorganism-to-culture medium ratio (10:90, 20:80, 30:70%), microbial source (pure *P. pentosaceus* vs. mixed TTAS consortium), and substrate (RSCB vs. HPSCB). In this pre-hydrolysis and SSF process, reducing sugar concentrations exceeding 40 g/L at 20 h using HPSCB or RSCB with *P. pentosaceus* at a 30:70 ratio. Significantly higher LA production with *P. pentosaceus* compared to TTAS: concentrations (11.51 g/L vs. 8.73 g/L), productivities (0.96 g/L·h vs. 0.18 g/L·h), and yields (0.57 g/g-SCB_{DW} vs. 0.43 g/g-SCB_{DW}) were achieved. This strategy demonstrates circular economy potential by valorizing SCB and anaerobic sludge (as carbon source and inoculum, respectively), reducing dependency on pure cultures/commercial enzymes while minimizing waste disposal. The approach provides a sustainable route for LA bioproduction and contributes to closing resource loops in the bioeconomy.

A. Santos Fuentefria et al. (Eds.): JIFREEEMA 2025, LNNS 1933, pp. 184–197, 2026.
https://doi.org/10.1007/978-3-032-23028-7_17

Keywords: heat-treated anaerobic sludge · hydrothermal pretreatment · lactic acid · *Pediococcus pentosaceus* · simultaneous saccharification and fermentation · sugarcane bagasse

1 Introduction

The concept of a circular economy and the waste-based biorefinery approach have led to the use of lignocellulosic biomass as a source of various value-added renewable chemicals. This currently constitutes one of the strategies to shift the paradigm from fossil-fuel-based chemical production to sustainable bio-based methods [1]. In this sense, sugarcane bagasse (SBC) generated in the sugar industry is a source of raw material with high availability and low cost that can contribute to the production of products with high utility value and costs.

Sugarcane bagasse is a lignocellulosic biomass derived after the extraction of sugarcane juice, which accounts for between 17 and 30% of the total sugarcane processed [2]. This biomass is rich in soluble sugars, cellulose (40–50%), hemicellulose (25–35%), lignin (10–30%), and ash (2,4%) [3, 4]. This composition enables its bioconversion to produce bioethanol, biogas, and other value-added bioproducts such as lactic acid (LA).

Lactic acid is a carboxylate with large market potential, with diverse applications such as the production of acrylic acid, biodegradable polymers and pyruvic acid, among others. LA can be synthesized as optically pure LA (D-LA or L-LA) or as a racemic mixture (DL-LA) by homofermentative and heterofermentative lactic acid bacteria strains. The specificity of LA enantiomers depends on the bacteria's lactate enzyme, which is responsible for catalyzing pyruvate to different LA enantiomers during lactate metabolism. L-lactate dehydrogenase converts pyruvate to L-lactate while D-lactate dehydrogenase generates D-lactate [5]. LA is a key intermediate bioproduct for the synthesis of biodegradable plastics such as polylactic acid (PLA) and other biodegradable chemicals. LA and PLA have wide applications in the food, pharmaceutical, textile, cosmetic, polymer, electronics, biomedical, agricultural, and packaging industries [4].

The production of LA from SBC is highly related to its composition, the highly polymerized phenolic lignin embedded in the polymeric matrix of hemicellulose and crystallized cellulose causes challenges in the conversion processes [6], preventing enzymes from successfully accessing and extracting fermentable sugars from cellulose [4]. Efficient reduction of biomass recalcitrance and recovery of the maximum carbohydrate fraction requires a pretreatment strategy [7]. There are several types of pretreatment for lignocellulosic materials: physical (mechanical and irradiation), chemical (alkaline, acidic, oxidizing, and organosolvent), physicochemical (extrusion, hydrothermal, and steam explosion), and biological (microorganisms and enzymatic), and their efficiency varies depending on the biomass and type of pretreatment [8].

The search for environmentally friendly pretreatments is a pressing need; therefore, hydrothermal pretreatment is one of the most promising. Pressurized hot water or hydrothermal pretreatment uses hot water at temperatures between 120 and 230 °C under high pressure. It consists of keeping the sugarcane bagasse in pressurized liquid hot water and promoting a rapid depressurization, after reaching a severity parameter, to break the lignocellulosic structure to provide the fermenting microorganism access to the cellulolytic substrate. However, at very high temperatures there is the possibility of formation of undesirable phenolic compounds and furan derivatives that can inhibit fermentation and therefore cause a decrease in the yield of methane, hydrogen or organic acids [9].

Simultaneous saccharification and fermentation (SSF) for the conversion of SCB into bioproducts offers several advantages. In this process, the saccharification of biomass and the fermentation of the resulting sugars occur simultaneously in a single bioreactor. Furthermore, the inhibition of the hydrolysis enzyme end product caused by the accumulation of glucose and cellobiose is minimized. This reduces capital and operating costs [10, 11].

In this sense, fungal fermentation with filamentous fungi such as *Aspergillus niger* has gained importance in numerous applications in the biotechnology industry due to its high production capacity, secretion efficiency, and ability to perform post-translational modifications, which favor the production of organic acids, proteins, and enzymes [12, 13]. *Aspergillus niger* is distinguished from other microbial cellular structures of bacterial origin by its ability to tolerate extreme culture conditions. It can be grown in a wide range of temperatures (10–50 °C), pH (2–11), salinity (0–34%), and in oligotrophic or nutrient-rich conditions. Significantly, this genus can degrade and utilize various biopolymers such as starch, hemicellulose, pectin, xylan, and proteins, allowing its cultivation on renewable resources such as plant biomass [12].

Current lactic acid bioproduction relies predominantly on pure microbial strains, requiring complex sterilization and nutrient supplementation. By contrast, the use of heat-treated anaerobic sludge (TTAS) as a mixed consortium offers operational and economic advantages, few LA fermentation studies have explored this source of inoculum, particularly for sugarcane bagasse valorization.

The objective of this work was to evaluate the feasibility of using thermally treated anaerobic sludge (TTAS) as an alternative microbial consortium to produce lactic acid (LA) from hydrothermally pretreated sugarcane bagasse (HPSCB) by comparing its performance against pure cultures such as *Pediococcus pentosaceus*.

2 Material and Methods

2.1 Substrate and Inoculum

Sugarcane bagasse (SCB) was obtained from the Boris Luis Santa Coloma Sugar Mill (Mayabeque, Cuba). The biomass was dried at 60 °C for 24 h, mechanically ground, and sieved to achieve a particle size < 2 mm. Processed SCB was stored in sealed nylon bags until use.

Anaerobic sludge was collected from Finca Marta (Artemisa, Cuba). For microbial activation, the sludge underwent heat treatment (90 °C, 15 min) following a modified protocol based on da Silva et al. [14].

Pediococcus pentosaceus LB-25 was provided by the Bank of Microorganisms for Animal Production (BAMIPA) at the Institute of Animal Science (ICA, San José de Las Lajas, Cuba). The strain was cultured in MRS medium [15] and maintained at 4 °C for experimental use.

2.2 Substrate Pretreatment

Hydrothermal pretreatment of SCB followed a 2^2 multifactorial design (temperatures: 120 °C/140 °C; residence times: 15/30 min) at 1:7 solid-to-liquid ratio (14% loading) [5]. The solids were then dried at 60 °C to reduce its moisture content and stored in air-tight plastic containers at room temperature until use.

2.3 Inoculum Preparation, Strains, and Culture Conditions

The heat-treated anaerobic sludge (TTAS) was prepared using a modified protocol of da Silva et al. [14]. First, the sludge was subjected to a heat shock treatment to inhibit methanogenic bacteria. Vacuum filtration was then performed to remove any coarse solids present. The pretreated and filtered inoculum was characterized. It was stored in a covered erlenmeyer flask and refrigerated until use. A 20 mL aliquot 10% (v/v) was then added to an erlenmeyer flask containing 180 mL of a *Clostridium*-Reinforced Medium (RCM), used for the selection and isolation of *Clostridium* species. To prepare the medium, the reagents were dissolved in distilled water to a final volume of one liter. The pH was adjusted to 6.8, and the mixture was autoclaved at 121 °C for 15 min. The modified RCM medium was prepared with the following composition: KH_2PO_4 (1 g/L), K_2HPO_4 (1 g/L), NaCl (5 g/L), $MnSO_4 \cdot 7H_2O$ (20 mg/L), $MgSO_4 \cdot 7H_2O$ (500 mg/L), $FeSO_4 \cdot 7H_2O$ (50 mg/L), CH_3COONH_4 (1.1 g/L), starch (1 g/L), agar (500 mg/L), peptone (10 g/L), yeast extract (3 g/L), meat extract (10 g/L), and glucose (10 g/L).

A strain of filamentous fungus *Aspergillus niger* from the Faculty of Chemical Engineering's culture collection was used as a microorganism with cellulolytic activity. It is preserved in malt agar wedges at 4 °C, using sugarcane bagasse for its potential as a producer of cellulase enzymes [16]. To obtain the enzymes, a spore solution is prepared as a preliminary step; for this, the microorganism strain was seeded in 500 ml erlenmeyer flasks containing the same agar medium used for conservation. The media were incubated for seven days at 30 °C to ensure sporulation of the microorganisms. Subsequently, 20 mL of sterile trace element solution was added to each erlenmeyer flask and placed on a shaker for 30 min at 140 rpm; then it was filtered through a sterile glass wool funnel, and the filtrate was collected in a sterile container. Finally, the number of spores is counted in a Neubauer chamber.

To ensure the nutritional requirements for the adequate development of microorganisms, the salts are dissolved in the established proportion KH_2PO_4 (4%), Urea (2%) and $(NH_4)_2SO_4$ (8.4%), then the pH is adjusted (4.5) and the amount of spore solution necessary to guarantee the humidity of the medium (70%) and the desired inoculum size (107 spores/$gSCB_{DW}$) is added. Subsequently, the bagasse and the solution are mixed manually for the necessary time until homogeneity is observed in the material. Fermentation was carried out in 500 ml erlenmeyer flasks for the time and temperature required for the adequate development of microorganisms. Subsequently, they are fermented for 72 h at 35 °C; after this time, the entire fermentation volume is pressed and then centrifuged to obtain the enzymatic preparation [16].

2.4 Enzymatic Hydrolysis

Enzymatic hydrolysis with *Aspergillus niger* enzymes followed a multifactorial experimental design $2^2 \cdot 4^1$ using raw (RSCB) and hydrothermally pretreated (HPSCB) sugarcane bagasse. The study evaluated temperatures (37 °C and 50 °C) and enzyme loadings (20, 40, 60, 80 FPU/g-SCB_{DW}) in 16 runs (one replicate). RSCB and HPSCB at 12% (w/w) was suspending in 100 mL of sodium-citrate buffer (pH 4.8). After pre-warming the suspension at 37 and 50 °C for 5 min, filter sterilized (through 0.2-μm membrane) cellulase was added at the designated loadings. The suspensions were then incubated at 37, pH 5,5 and 50 °C for 72 h. Samples were taken every 24 h during incubation. The samples were centrifuged at 10,000 rpm for 10 min, and the supernatants were collected for the determination of reducing sugar concentration (DNS method [17]) served as the response variable.

2.5 Lactic Acid Production via Pre-hydrolysis Followed by SSF

Optimal LA production conditions were identified via a multifactorial experimental design $2^3 \cdot 3^1$ (24 runs, two replicates), evaluating the inoculum-to-culture medium ratios (10–90, 20–80, 30–70% v/v), type of inoculum (*P. pentosaceus* vs. TTAS consortium), substrate (RSCB vs. HPSCB), and fermentation time (24/60 h). Fixed conditions of this process were a 10% w/v solid loading, 150 rpm, pH 6.5, 37 °C, 250 mL working volume [10]. Following the pre-saccharification procedure of Haokok et al. [10], the RCM was first inoculated with *Aspergillus niger* (microorganism with cellulolytic activity) at 50 °C, 150 rpm for 12 h. After this time, for SSF initiation the TTAS or *P. pentosaceus* were added and incubated, decreasing the temperature to 37 °C, (optimal for lactic acid bacteria) [18, 19].

2.6 Analytical Methods

The physicochemical characterization of the sugarcane bagasse and the anaerobic inoculum included the analysis of total solids (TS), volatile solids (VS), and ash, according to standardized methods [20]. In addition, the composition of cellulose, hemicellulose, and lignin in SCB (treated and untreated) was determined according to the standards: ANSI/ASTM D1103-60, ANSI/ASTM D1106-56 and ANSI/ASTM D1105-56, respectively [21]. The pH was determined using a CRISSON PH 25 pH meter.

The lignin removal efficiency, as well as the solubilization of hemicellulose and cellulose, after pretreatment was calculated based on the mass of the component in the sugarcane bagasse before and after the pretreatment, using Eq. 1 [10].

$$R = \left(1 - \frac{m}{m_0}\right) \cdot 100 \qquad (1)$$

R: efficiency
m: mass of the component in the pretreated bagasse
m_0: mass of the component in the untreated bagasse

The reducing sugar content was determined by spectrophotometry with measurement at 550 nm, using the Miller method with 3,5-dinitrosalicylic acid or DNS according to the procedure described by Morales et al., [17]. Lactic acid content was determined spectrophotometrically at 390 nm, following a procedure described by [22].

3 Results and Discussion

3.1 Characterization of Sugarcane Bagasse

The sieved bagasse samples were determined for total solids (TS), total volatile solids (TVS) and ash content according to standardized methods [20]. Lignin, hemicellulose and cellulose contents were also determined based on [21]. The results of the characterization of the RSCB showed a high content of total volatile solids (95%), making it a material with the potential to be biodegraded anaerobically. Previous studies, González et al., [21] and Cao et al., [23] have reported similar values of volatile solids for sugarcane bagasse, in the range of 96–98%. Regarding the total solids content (96,2%), another author [24] reported that a lower amount of these and a higher water content could provide a homogeneous environment and improve the accessibility of the microorganism to the substrate because of a higher conversion to product. The pH of sugarcane bagasse (4.16) is usually in the range of 4 to 7, which classifies it as slightly acidic to neutral. However, it is important to note that this value may vary depending on different factors such as the type of cane, the cultivation conditions and the juice extraction process [25].

The results regarding the lignin, cellulose and hemicellulose content (13,32; 36,4 and 8,36% respectively) of the RSCB showed differences to those reported in studies carried out by García et al., [26], which reached values of 15,72; 52,98 and 26,53% respectively. On the other hand, Avellán et al.,[27] in the characterization of the bagasse obtained values of 18,9% lignin, 68,37% cellulose and 3,06% hemicellulose, with similar results to the ash content (3,24%). Also, in the characterization carried out by Haokok et al., [10] cellulose contents of 39.8% and hemicellulose 19.5% are reported, values close to those of the present study. These differences may be due to the variety or origin of the raw material used, with respect to the difference in cellulose content is very marked, which leads to infer that the variety may be the phenomenon that justifies these results [28]. The discrepancy between the characterization of the sugarcane bagasse used with respect to studies conducted by other authors, could be influenced by genetic, climatic, soil quality, growth area, agricultural management and variety factors [29].

3.2 Hydrothermal Pretreatment

Hydrothermal pretreatment was studied for sugarcane bagasse pretreatment, as it is one of the most environmentally friendly pretreatments due to the use of mild conditions and the lack of reagents. In this pretreatment, the sugarcane bagasse is subjected to two temperature levels (120 and 140 °C) and two contact times (15 and 30 min) (Table 1). Hydrothermal pretreatment carried out at 140 °C for 15 min, was the most promising case studied, as the lignin content was reduced to 5.02%, corresponding to 62,31% delignification. This represents more than half of the initial lignin content, indicating an optimal point for the process under this condition.

Table 1. Lignin content of hydrothermally pretreated sugarcane bagasse

Temperature (°C)	Time (min)	Lignin (%)	Delignification (%)
120	15	11.96	10.93
120	30	7.12	46.54
140	15	5.02	62.31
140	30	7.10	46.69

These results are consistent with those achieved by López et al. [30], who also obtained the best pretreatment conditions at 140 °C and 12.5 min. Similar temperatures and times have also been reported by García et al. [26], who achieved their best condition at 121 °C for 60 min. Although they did not base their results on the percentage of delignification, they did base their results on the concentration of glucose released. Changes in the composition of sugarcane bagasse after hydrothermal pretreatment were evaluated under four different conditions reported in Table 2.

Table 2. Physicochemical characterization of pretreated sugarcane bagasse

Parameters	Bagasse 120 °C 15 min	Bagasse 120 °C 30 min	Bagasse 140 °C 15 min	Bagasse 140 °C 30 min
TS (%)	95.72	95.60	94.46	94.09
TSV (%)	77.12	71.56	70.13	69.87
Ashes (%)	2.65	2.67	2.92	2.91
Lignin (%)	11.96	7.12	5.02	7.10
Cellulose (%)	46.0	42.0	60.8	55.5
Hemicellulose (%)	10.4	11.45	11.80	10.12
pH meter	3.87	4.21	4.35	4.16
Conductivity (μS/cm)	517	384	428	436

Similar values to those achieved in bagasse pretreatment have been reported by several authors [10, 31]. As can be seen in the table above, while lignin content decreases, cellulose and hemicellulose content increases. This is the result of the fact that when lignin is separated from cellulose and hemicellulose, a greater amount of these polysaccharides are available in forms that are more accessible for subsequent processing, such as enzymatic hydrolysis [32]. In a similar investigation with SCB, Vera et al. [6] reported lignin, cellulose, and hemicellulose values of 11,59; 51,26 and 25,18% respectively, while [10] the alkaline-treated SCB the values were 7,3% lignin, 66,8% for cellulose and 13,9% for hemicellulose.

The results obtained in this research are in line with the values reported for the lignocellulosic composition of sugarcane bagasse. The similarity between these values suggests that the pretreatment conditions and the inherent characteristics of sugarcane bagasse maintain a relative uniformity between different samples and analysis methods. After pretreatment, there were no significant changes in terms of total solids, not so for the volatile solids content where a decrease is observed with respect to untreated bagasse, which is attributed to the fact that some of these components that are more resistant to solubilization remained in the hydrolyzate [21]. This hydrothermal pretreatment did not produce significant changes in the pH and conductivity of the sugarcane bagasse.

3.3 Inoculum Characterization

The anaerobic sludge used as inoculum was subjected to heat shock treatment at 95 °C for 30 min to inhibit methanogenic bacteria. Non-spore-forming bacteria such as Clostridium and bacteria that produce lactic, acetic, and propionic acids are generally resistant, whereas hydrogenotrophic methanogens such as *Enterobacter or Citrobacter*, which generate hydrogen, are eliminated [33]. The results show that heat shock pretreatment and vacuum filtration decreased the concentrations of total, fixed, and volatile solids from (33,454 to 1,915), (15,270 to 1,060), and (18,184 to 760), respectively. This is mainly due to the filtration process. Therefore, it can be stated that the STV content is adequate to ensure an appropriate concentration of microbial biomass at the start of fermentation [21]. On the other hand, it did not produce significant changes in the pH (7.81 to 7.88) and conductivity (4.92 to 4.29) of the anaerobic sludge.

3.4 Enzymatic Hydrolysis

After the pretreatment was completed, enzymatic hydrolysis was carried out to convert the cellulose into glucose without the formation of inhibitory compounds. The effect of bagasse pretreatment, temperature, and enzyme load were evaluated during the enzymatic hydrolysis. Figure 1 shows the behavior of reducing sugar concentration during enzymatic hydrolysis using RSCB and HPSCB with different enzyme loads at two different temperatures (Fig. 2).

At a temperature of 50 °C and an enzyme load of 80 PFU/g, the maximum reducing sugar concentration was achieved at 24 h, corresponding to sample RSCB, with a value of 11,98 g/L. Estrada [34], obtained a reducing sugar concentration of 9,03 g/L, but with an incubation period of 12 h. The results obtained in the characterization of SCB contrast with similar studies conducted with the same residue. Ferrer et al. [35] obtained reducing sugars of 6,09 g/L. The variability in the levels of reducing and extractable sugars may be due to both the intrinsic heterogeneity of the bagasse and differences in the extraction methods used in each study.

The effect of substrate type, enzyme load, and temperature on reducing sugar concentration was investigated using a $2^2 . 4^1$ multifactorial experimental design with one replicate, resulting in 32 runs. The statistical analysis showed that four effects (temperature, type of substrate, combination of temperature and substrate and combination of load) are statistically significant with 95% confidence for the production of lactic acid because they present P-values < 0.05.

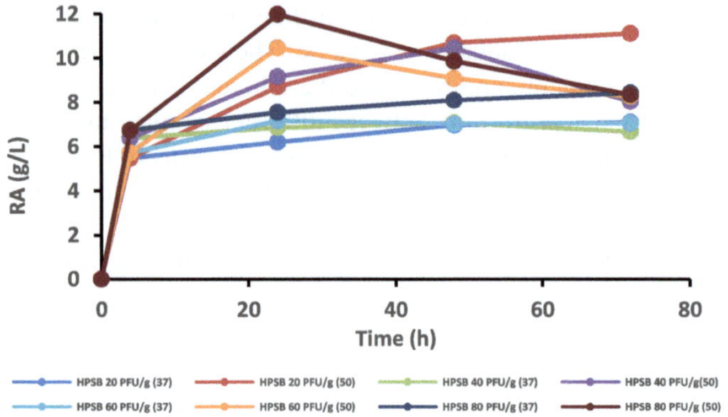

Fig. 1. Concentration of reducing sugars for enzymatic hydrolysis performed on pretreated bagasse

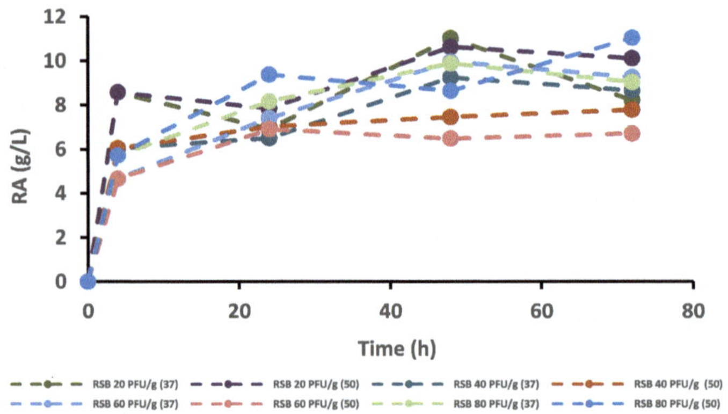

Fig. 2. Concentration of reducing sugars for enzymatic hydrolysis performed of raw sugarcane bagasse.

3.5 Pre-hydrolysis Followed by SSF

Simultaneous saccharification and fermentation (SSF) is achieved by adding the required hydrolytic enzymes to the fermentation broth, followed by inoculation of the microorganism, allowing it to ferment the sugars as they are released from the biomass by enzymatic action [31]. SSF has the advantage of simplifying the process, as it prevents the accumulation of fermentable sugars, which are transformed by the microorganisms immediately after their release. It also prevents product inhibition and increases the overall yield and efficiency of the process. It also allows for a reduction in processing times by carrying out enzymatic saccharification and fermentation simultaneously. This reduction in processing time allows for increased installed plant capacity, improving productivity and reducing production costs [36].

For both inoculums source, *Pediococcus pentosaceus* and thermally treated anaerobic sludge, ratios of 10–90, 20–80, and 30–70% with the culture medium were used. Both were incubated at 37 °C for 48 h, followed by fermentation. To increase lactic acid production, fermentation was carried out under the conditions described above. The effects of substrate pretreatment, inoculum type, and medium-to-inoculum ratio were evaluated. Figures 3 and 4 show how the concentration of lactic acid generated during fermentation varies when different types of inoculums source are used and the proportion between the culture medium and the inoculum is modified.

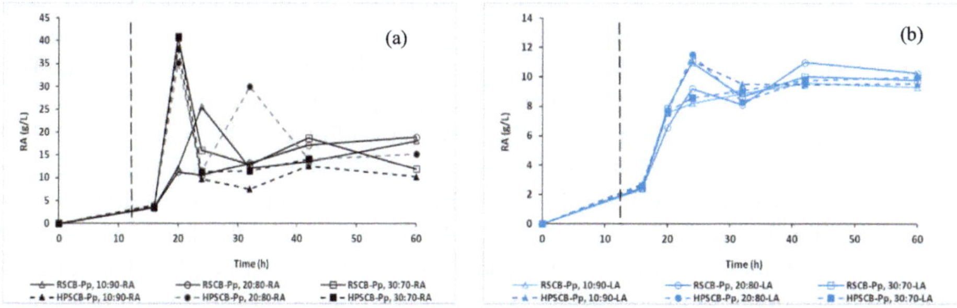

Fig. 3. Fermentation kinetics of (a) reducing sugar consumption and (b) LA production using *Pediococcus pentosaceus*

Figures 3(a and b) illustrate the fermentation kinetics of reducing sugar consumption and lactic acid (LA) production under varying inoculum conditions. The highest LA titers (11.51 g/L) and substrate conversion rates (0.96 g/L·h) were achieved *P. pentosaceus* at 20:80 medium-to-inoculum ratio (Fig. 3b), coinciding with rapid sugar depletion within 20h (Fig. 3a).

In contrast, TTAS consortia (Fig. 4a and b) showed prolonged sugar utilization (>40h) and lower LA accumulation (8.73 g/L), indicating microbial composition critically governs process efficiency.

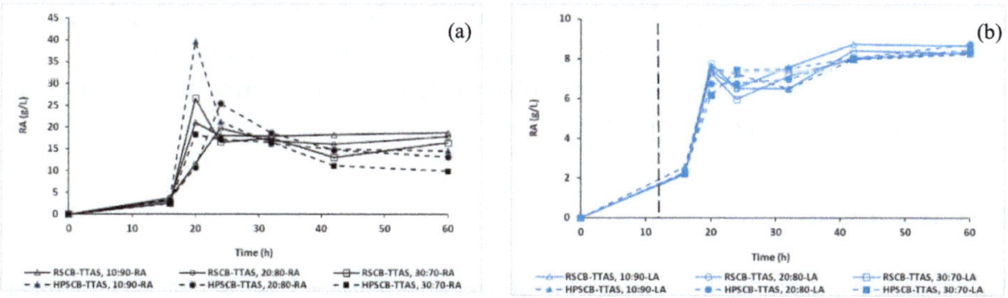

Fig. 4. Fermentation kinetics of (a) reducing sugar consumption and (b) LA production using TTAS.

In contrast, TTAS consortia (Fig. 4a and b) showed prolonged sugar utilization (>40h) and lower LA accumulation, indicating microbial composition critically governs process efficiency. A maximum concentration of lactic acid is obtained at 60 h (8,73 g/L) for variant HPSCB-TTAS, which contains an inoculum-medium ratio of 20–80%, whose productivity and lactic acid yield values for this sample are 0,18 g/Lh and

0,43 g/g respectively. It can be highlighted that between both inoculums source, the best results are obtained with the pure culture, since, during the fermentation process with mixed cultures, the desired product is obtained along with other unwanted fermentation products [33]. This is an inherent disadvantage that causes a decrease in yield and productivity.

Comparable lactic acid concentrations were observed between this study (11.51 g/L) and prior works, Sornlek et al. [37] reported 10.24 g/L using sugarcane bagasse with *Saccharomyces cerevisiae*, while Ahmad et al. [24] achieved 10.9 g/L with *Lactobacillus lactis*. These values are substantially lower than the 86.7 g/L reported by Haokok [10], where alkaline pretreatment achieved near-complete lignin removal (89% delignification), a critical factor enhancing cellulose accessibility. In our system, Figs. 3 and 4 reveal characteristic LA kinetic profiles: rapid concentration increase to a maximum (20 h for *P. pentosaceus*; 36 h for TTAS), followed by a gradual decline attributed to product inhibition or microbial decay. This behavior reflects the fundamental stages of a dark fermentation process using sugarcane bagasse as a substrate [38]. The samples analyzed in the previous figures, which contain a maximum concentration of lactic acid and were previously pretreated, demonstrate the effect of hydrothermal pretreatment on sugarcane bagasse, which is one of the objectives of this research.

The design of experiments assessed the impact of substrate pretreatment (raw vs. hydrothermally treated SCB), inoculum type (pure *P. pentosaceus* vs. TTAS consortium), and inoculum-to-medium ratio (10–30% v/v) on lactic acid production. The model demonstrated strong predictive capability ($R^2 = 0.9057$), with ANOVA identifying three statistically significant factors ($p < 0.05$) at 95% confidence: fermentation time, substrate type, and inoculum-time interaction. The maximum LA concentration (11.51 g/L) was achieved at 24 h using pretreated bagasse with *P. pentosaceus*, consistent with kinetic optimization patterns observed in Figs. 3 a and b.

This study establishes a sustainable pathway for lactic acid production through dual valorization of agro-industrial residues: sugarcane bagasse as carbon source and anaerobic sludge as microbial inoculum. The integrated prehydrolysis-SSF approach using TTAS represents a circular economy innovation, reducing dependency on pure cultures while valorizing waste streams. This biorefinery model exemplifies how synergistic waste valorization can advance biobased chemical production while addressing agricultural waste management challenges.

4 Conclusions

This study demonstrates the viability of valorizing sugarcane bagasse (SCB) and anaerobic sludge for sustainable lactic acid (LA) production via an integrated prehydrolysis-SSF platform. The main results reveal that hydrothermal pretreatment (140 °C, 15 min) enhances SCB digestibility (62.3% delignification), yielding 11.98 g/L reducing sugars during enzymatic hydrolysis. Pure *P. pentosaceus* outperformed TTAS consortia, yielding 32% higher lactic acid (LA) concentrations (11.51 g/L vs. 8.73 g/L) and greater productivity (0.96 g/L·h vs. 0.18 g/L·h). Nevertheless, TTAS remains a viable low-cost inoculum, demonstrating the feasibility of waste-derived microbial resources.

This work establishes a scalable prehydrolysis-SSF platform for sustainable lactic acid production, transforming sugarcane bagasse and anaerobic sludge from waste streams into valuable bio-based products.

References

1. Lee, J.A., Kim, J.Y., Ahn, J.H., Ahn, Y.J., Lee, S.Y.: Current advancements in the bio-based production of polyamides. Trends Chem. **5**(12), 873–891 (2023)
2. Castillo Alvarez, Y., et al.: Transition to a circular bioeconomy in the sugar agro-industry: predictive modeling to estimate the energy potential of by-products. Technologies **13**(6), 238 (2025)
3. Talha, Z., Ding, W., Mehryar, E., Hassan, M., Bi, J.: Alkaline pretreatment of sugarcane bagasse and filter mud codigested to improve biomethane production. Biomed. Res. Int. **2016**(1), 8650597 (2016)
4. Agrawal, D., Kumar, V.: Recent progress on sugarcane-bagasse based lactic acid production: technical advancements, potential and limitations. Ind. Crops Prod. **193**, 116132 (2023). https://doi.org/10.1016/j.indcrop.2022.116132
5. Ajala, E.O., Olonade, Y.O., Ajala, M.A., Akinpelu, G.S.: Lactic acid production from lignocellulose–a review of major challenges and selected solutions. ChemBioEng Rev. **7**(2), 38–49 (2020). https://doi.org/10.1002/cben.201900018
6. Vera Dueñas, M.O., Romero Moreira, A.J., Zambrano Zambrano, G.A., Ponce, W., Riera, M.A.: El bagazo de la caña de azúcar como sustrato en la fermentación láctica. Avances en Química **19**(1), 27–36 (2024)
7. Vele Salto, A.M., Abril González, M.F., Zalamea Piedra, T.S., Pinos Vélez, V.P.: Mini revisión: aplicación de líquidos iónicos en hidrólisis ácida de material lignocelulósico para la obtención de azúcares. Ciencia en desarrollo **12**(1), 55–67 (2021). https://doi.org/10.19053/01217488.v12.n1.2021.12477
8. Du, J., Qian, Y., Xi, Y., Lü, X.: Hydrothermal and alkaline thermal pretreatment at mild temperature in solid state for physicochemical properties and biogas production from anaerobic digestion of rice straw. Renew. Energy **139**(C), 261–267 (2019). https://doi.org/10.1016/j.renene.2019.01.097
9. Talaiekhozani, A., Rezania, S.: A critical review on the various pretreatment technologies of lignocellulosic materials. J. Environ. Treat. Tech. **8**(3), 925–935 (2020)
10. Haokok, C., Lunprom, S., Reungsang, A., Salakkam, A.: Efficient production of lactic acid from cellulose and xylan in sugarcane bagasse by newly isolated Lactiplantibacillus plantarum and Levilactobacillus brevis through simultaneous saccharification and co-fermentation process. Heliyon. **4**(7), 9 (2023). https://doi.org/10.1016/j.heliyon.2023.e17935
11. Abdel-Rahman, M.A., Tashiro, Y., Sonomoto, K.: Recent advances in lactic acid production by microbial fermentation processes. Biotechnol. Adv. **31**(6), 877–902 (2013). https://doi.org/10.1016/j.biotechadv.2013.04.002
12. Kurt, T., Marbà-Ardébol, A.M., Turan, Z., et al.: Rocking *Aspergillus*: morphology-controlled cultivation of *Aspergillus niger* in a wave-mixed bioreactor for the production of secondary metabolites. Microb. Cell Fact. **17**, 128 (2018). https://doi.org/10.1186/s12934-018-0975-y
13. de Figueiredo, F.L., et al.: Ingeniería del secretoma de *Aspergillus niger* para la producción de celooligosacáridos a partir de biomasa vegetal. Fábricas de células microbianas **23**(1), 323 (2024). https://doi.org/10.1186/s12934-024-02578-9
14. da Silva, D.B., Fernandes, B.S., da Silva, A.J.: Effect of initial pH and substrate concentration on the lactic acid production from cassava wastewater fermentation by an enriched culture of acidogenic microorganisms. Water Environ. Res. **93**(10), 1925–1933 (2021). https://doi.org/10.1002/wer.1467

15. García-Hernández, Y., Pérez-Sánchez, T., García-Curbelo, Y., Sosa-Cossio, D., Nicoli, J.R.: Growth ability, microbial activity and susceptibility to antimicrobials of two strains of Pediococcus pentosaceus, candidates to probiotic. Cuban J. Agric. Sci. **51**(4), 433–442 (2017)
16. Dustet, J., Carmenate, M., Haramboure, T., Hernández, O., Martínez, J.L.: Obtención de celulasas de Aspergillus niger, técnica de cultivo y método de separación. Cuba. Pág, 1–50 (2000)
17. Morales, A.J., Vivas, Y.A.: Evaluación de la actividad antioxidante de una bebida refrescante a base de lactosuero adicionada con pulpa de curuba (Passiflora Mollissima Bailey), durante su almacenamiento (2015). Recuperado de: https://hdl.handle.net/20.500.14625/32848
18. Espinosa-Negrín, A.M., López-González, L.M., Casdelo-Gutiérrez, N.L.: Pretratamientos aplicados a biomasas lignocelulósicas: una revisión de los principales métodos analíticos utilizados para su evaluación. Revista Cubana de Química **34**(1), 87–110 (2022)
19. Palomar, L.T., et al.: Determinación simplificada del contenido de Lignina en paja de trigo por método volumétrico y su caracterización química. Cienc. Tecn. **15**, 113–124 (2015)
20. APHA-AWWA-WEF: Standard Methods for Examination of Water and Wastewater. American Public Health Association, Washington DC, USA (2012)
21. González, A., Hernández, G., Pereda, I.: Pretratamiento alcalino de bagazo de caña para mejorar la producción de biometano. Centro Azúcar **46**(4), 79–88 (2019)
22. Borshchevskaya, L.N., Gordeeva, T.L., Kalinina, A.N., Sineokii, S.P.: Spectrophotometric determination of lactic acid. J. Anal. Chem. **71**(8), 755–758 (2016). https://doi.org/10.1134/S1061934816080037
23. Cao, W., Sun, C., Li, X., Qiu, J., Liu, R.: Methane production enhancement from products of alkaline hydrogen peroxide pretreated sweet sorghum bagasse. RSC Adv. **7**(10), 5701–5707 (2017). https://doi.org/10.1039/c6ra25798d
24. Ahmad, A., Banat, F., Taher, H.: Enhanced lactic acid production from food waste in dark fermentation with indigenous microbiota. Biomass Conv. Bioref. **12**, 3425–3434 (2022). https://doi.org/10.1007/s13399-020-00801-2
25. Culqui, R.G., Pilco, C.J., Narváez, L.G., Guangasi, T.M.: Revalorización del bagazo de caña de azúcar (*Saccharum officinarum*) como residuo importante para la agroindustria. 593 Digital Publisher CEIT **8**(3), 134–148 (2023). https://doi.org/10.33386/593dp.2023.3.1661
26. García, E., Peralta, D.: Hidrólisis enzimática de bagazo de caña (Saccharum sp hibrido) para la producción de azúcares simples. Agroindustria, Sociedad y Ambiente **1**(18), 119–131 (2022). https://doi.org/10.5281/zenodo.6481849
27. Avellan Rengifo, A.C., Mendoza Palma, A.N., Ponce, W., Riera, M.A.: Sugarcane bagasse as a material in the production of polylactic acid. Int. J. Ind. Chem. **15**(3), 1–11 (2024). https://doi.org/10.57647/j.ijic.2024.1503.17
28. Esquivel-Hernández, D.A., García-Pérez, J.S., López-Pacheco, I.Y., Iqbal, H.M., Parra-Saldívar, R.: Resource recovery of lignocellulosic biomass waste into lactic acid-trends to sustain cleaner production. J. Environ. Manage. **301**, 113925 (2022). https://doi.org/10.1016/j.jenvman.2021.113925
29. Resano, D., Guillen, O.W., Ubillús, F.D., Barranzuela, J.L.: Caracterización fisicoquímica del bagazo de caña de azúcar industrial y artesanal como material de construcción. Información tecnológica **33**(2), 247–258 (2022). https://doi.org/10.4067/S0718-07642022000200247
30. López González, L.M., Pereda Reyes, I., Dewulf, J., Budde, J., Heiermann, M., Vervaeren, H.: Effect of liquid hot water pre-treatment on sugarcane press mud methane yield. Biores. Technol. **169**, 284–290 (2014). https://doi.org/10.1016/j.biortech.2014.06.107
31. Vegas Niño, R.M., Acosta Arteaga, Y., Fernández Segura, C.: Obtención, purificación y aprovechamiento de azúcares reductores a partir de materiales lignocelulósicos. Rev. Bol. Quim **41**, 43–60 (2024). https://doi.org/10.34098/2078-3949.41.2.2
32. Ortiz, W.G.C.: Tratamientos aplicables a materiales lignocelulósicos para la obtención de etanol y productos químicos. Revista de Tecnología (Archivo) **13**(1), 39–44 (2014)

33. Dareioti, M.A., Vavouraki, A.I., Tsigkou, K., Zafiri, C., Kornaros, M.: Dark fermentation of sweet sorghum stalks, cheese whey and cow manure mixture: effect of pH, pretreatment and organic load. Processes **9**(6), 1017 (2021). https://doi.org/10.3390/pr9061017
34. Estrada-Maya, A., Weber, B.: Biogás y bioetanol a partir de bagazo de agave sometido a explosión de vapor e hidrólisis enzimática. Ingeniería investigación y tecnología **23**(2), 1 (2022). https://doi.org/10.22201/fi.25940732e.2022.23.2.009
35. Ferrer, J.R., Páez, G., Arenas de Moreno, L., Chandler, C., Mármol, Z., Sandoval, L.: Cinética de la hidrólisis ácida de bagacillo de caña de azúcar. Rev. Fac. Agron. **19**(1), 23–33 (2002)
36. Proaños, J., Piñeros Castro, Y.: Evaluación de la producción de ácido láctico a partir de cascarilla de arroz por *Lactobacillus delbrueckii*. Revista Mutis **4**(1), 33–39 (2014). https://doi.org/10.21789/22561498.908
37. Sornlek, W., et al.: D-Lactic acid production from sugarcane bagasse by genetically engineered *Saccharomyces cerevisiae*. J. Fungi **8**(8), 816 (2022). https://doi.org/10.3390/jof8080816
38. Asunis, F., et al.: Integrating treatment processes for cheese whey valorisation. J. Environ. Manag. **276**, 111240 (2020). https://doi.org/10.1016/j.jenvman.2020.111240

Competitiveness Factors of Cuban Sugarcane Value Chain

Arodis Caballero Núñez[1](✉), Ester Susana Michelena Fernández[2], and Ariel Racet Valdés[2]

[1] Instituto Cubano de Investigaciones de los Derivados de La Caña de Azúcar, La Habana, Cuba
arodis.caballero@icidca.azcuba.cu

[2] Universidad Tecnológica de La Habana "José Antonio Echeverría", La Habana, Cuba

Abstract. The factors that differentiate the management of 38 Cuban sugar companies in the period 2020–2024 are identified and evaluated, characterized by the decrease in the area covered with cane, cessation of exports and a sharp deterioration of the working and living conditions of workers in the sector, although positive elements are also present such as a different look at local chains, agricultural diversification with food production and the need for innovation and the introduction of new scientific results. Based on strategic foresight studies and diagnoses in sugar companies and cooperatives, factors of change (variables) associated with competitiveness were identified and a competitiveness indicator is designed and evaluated, which allows the elaboration for the first time of a ranking where gaps, deficiencies or weaknesses can be quickly and effectively identified for decision-making (investments, systematic evaluations, continuous improvement).

Keywords: competitiveness · value chain · sugar · factor analysis

1 Introduction

The Cuban sugar agroindustry has historically played a key role in the country's exports and domestic food production, but sugar exports ceased in 2023. The sharp decline in production levels of sugarcane and its main by-products is placing the sector in a major crisis, the causes of which are varied, both external and internal. On the one hand, the impact of the aggressive US financial and trade embargo and, on the other, the lack of effective linkages and sectoral policies have led to a loss of competitiveness in its business management and a notable vulnerability of its financial, logistical, and technological systems. The most notable impact is the worse use of land and very low exploitation of its aging infrastructure (machinery, irrigation, workshops).

The sugarcane value chain is a complex set of relationships, processes and actors whose results are evaluated using a large number of indicators, also useful in the preparation of forecasts, planning and classification actions, among others [1].

Efficient management of the chain ensures the sustainability of this type of agribusiness, which is oriented in three dimensions: guarantee of economic return, social development and environmental responsibility, considering compliance with regulations, return

A. Santos Fuentefria et al. (Eds.): JIFREEEMA 2025, LNNS 1933, pp. 198–208, 2026.
https://doi.org/10.1007/978-3-032-23028-7_18

on investment, competitive advantages in the market and reputation and image [2–4]. Other authors place innovation, productive flexibility and quality as key elements [5, 6].

Five key processes are constituents of sugarcane value chain: agricultural production, harvesting and transportation, production of sugar and by-products, production of derivatives and marketing. Their input and output indicators can be classified in many ways, depending mainly on their purpose of use, such as efficiency, productivity, quality, economic, human capital, social, environmental and innovation and we can locate and evaluate them at the entrance or exit of the stages. For this, in some cases, continuous physical measurements are available (metrological, records, economic calculations or any other type) and in other cases the estimation of the variable is done discontinuously (yes or no, much or little) through more subjective assessments through surveys or other tools with the participation of experts, managers, technicians, workers and stakeholders.

This study aims to identify and evaluate the factors that most impact the competitiveness of Cuban sugar companies today.

An useful indicator is proposed for decision-making, the systematic evaluation of sugar companies, and the identification of gaps that must be addressed to achieve sustainable results and capital growth.

2 Methods

A comprehensive documentary study was conducted with recent information related to sugarcane production and the efficiency of sugar and sugar derivatives production in sugar companies and production bases in Cuba and other countries. The analysis period includes two disruptive moments: 2002, when the sugar sector underwent a disruptive resizing, and 2020, due to the impact of the implementation of the so-called Ordering Task on the costs, prices, and economic results of both sugarcane cooperatives and sugar companies. The indicators analyzed are from subsequent to both events, as they better characterize the current situation.

Strategic foresight studies were carried out in sugar companies following Godet 's methodology with important contributions from other authors [7, 8]. There were applied surveys in sugarcane cooperatives in order to identify the factors of change and elements of the value chain with which a new management model based on innovation and focused on competitiveness could be designed.

Mixed methods were used for data analysis. Various questionnaires, surveys, interviews, and other procedures were designed and processed to evaluate the input and output variables of key processes. The qualitative variables used in the research were parameterized. The majority of the quantitative variables were organized and standardized.

For the processing of the information obtained, three basic applications were used: MIC-MAC for strategic foresight studies, the different applications contained in Microsoft Excel version 2013 for storage and preparation of databases and multivariate analysis techniques using Statgrahics. Centurion XVI for factor analysis.

Multivariate factor analysis techniques allowed us to dimension hidden variables in this information, such as competitiveness, a quality of the organization that required a preliminary definition [9].

The following glossary explains the meaning of some terms used in factor analysis:

1. Eigenvalue also known as characteristic value is a scalar used to represent the magnitude of the transformation of vector. Those factors with eigenvalue > 1 are selected for processing.
2. Factor loading indicates the amount of variance that a feature has in common with the total pool of shared variance and ranges from 0 to '/– 1. It is used in Eq. (2) to calculate ICOM index by using factors.

3 Results

By making foresight studies conducted at 54 sugar companies more than 100 internal and external factors were identified, with a high degree of repetition due to the characteristics of Cuban companies (high centralization, homogeneity in their structure, uniformity in governance forms).

Seventeen internal weaknesses were associated with the most common definitions of the concept of competitiveness. Five of them were related to supply aspects (low sugarcane production, difficulties in accessing external financing, low investment, weak supply chain and non-reproduction of working capital); two to the workforce (its fluctuation and failures in training and motivation), three to administration (lack of innovation, efficient management models and selection and preparation of managers); three to technological aspects (obsolescence, failures in maintenance systems, in operation) and two to market aspects (quality, failures in chains).

Furthermore, the diagnoses carried out on 645 cooperatives made it possible to identify the relationship between the poor results in sugarcane production and the instability of basic supplies (fuel, fertilizers), migration, and other phenomena that are closely associated with the deterioration of rurality in the environment of the Cuban sugarcane farming.

A set of results of 90 farmers who exceeded 54 t/ha (a target set in the sector's restructuring process) and produced more than 1.4 million t of sugarcane in the five-year period were analyzed. They maintained high sugarcane yields despite the multidimensional crisis. The improved economic and productive management of these producers included indicators such as the elimination of idle sugarcane areas, the quality of agricultural work, the use of crop rotation and intercropping for food production to meet community demand and increase incomes, industrialization, the use of bioproducts, and other technological advances in cultivation and harvesting, including advanced methods such as satellite georeferencing [10].

Two new indicators were obtained: a diversification index that assesses the degree to which new products and technologies are introduced, and a dimensionless, multicomponent competitiveness index for the Cuban sugar industry.

Due to the lack of specific measurements, it is assumed that the competitiveness of Cuban sugar companies during the analysis period (2020–2024) would be decidedly low due to poor production results. However, partial evaluations allow us to confirm—as in the case of sugarcane producers—that some companies can show factors that positively boost competitiveness both in marketing and in the rest of the key processes involved in the production of sugar and sugar products.

Prior to its mathematical formulation, a verbal definition of competitiveness appropriate to the conditions of the Cuban sugar sector was made.

A sugar company or sugarcane cooperative is competitive if it is efficient, productive, offers quality products and services, is profitable, has trained, stable, and committed workers, and is also innovative and embraces its social responsibility, which includes environmental respect and all kinds of support for the communities where its production operations are carried out.

The study included the update of the sugarcane value chain, which allowed for the identification and evaluation of more than one hundred indicators that systemically measure the performance of the variables previously identified in the foresight exercises and those contained in the previous verbal definition [11].

For practical reasons, the number of indicators for each key process and dimension of the management model had to be reduced without this reduction process jeopardizing the ability to measure the impacts that each of them has on the overall result.

Two elements were used for this: (1) expert judgment, which determined the dependence and importance of each of the selected indicators, and (2) factor analysis, which grouped their specific weight in the final result. A summary of this work is provided in Fig. 1.

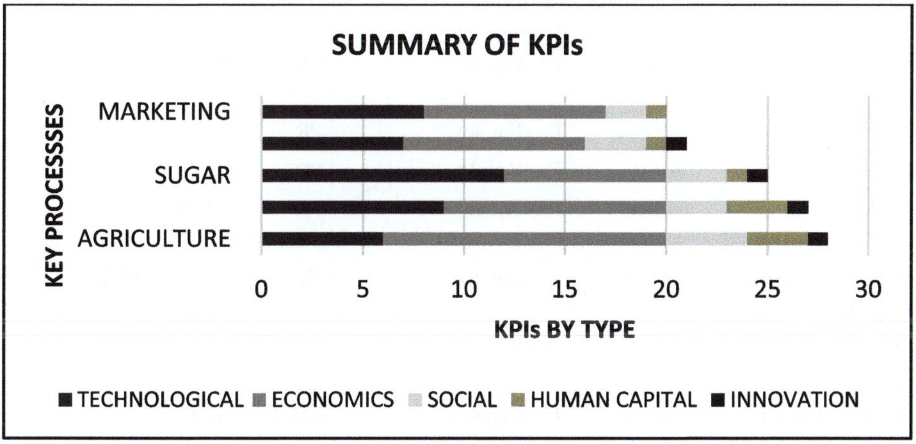

Fig. 1. Summary of key performance indicators by processes. Source: elaboration own.

A study was conducted on the efficiency and productivity of sugar companies during this period. Indicators such as capacity utilization, industrial efficiency, potential sugarcane yield utilization (PCR), sugarcane recovery, and others are systematically assessed to evaluate industrial efficiency losses. In distilleries, criteria such as fermentation and distillation efficiency, specific molasses consumption per hectoliter of alcohol, and others are also included.

To measure productivity and quality, the area covered by sugarcane or food, the quantity of sugarcane produced per hectare, the production of sugarcane per farm worker, the sugar content in the sugarcane, the indicators that the client demands in the quality of the products (humidity, iron or dextran content, to name only the negative ones), the alcoholic strength and the organoleptic indicators of the rum were taken into account.

The big data obtained through the study includes historical series and trend analyses of these indicators and was enriched with other indicators associated with the social impact of the activities of companies and cooperatives, such as wages or economic contributions to the development of communities and bateyes.

The indicators selected for the model are listed in Table 1.

Table 1. List of key performance indicators by type, dimension, and associated processes. (n = 16)

Name of the indicator	Code	Dimension	Guy	Associated processes
Industrial performance	RIND	Prod/quality	EF	AZ
Utilization of potential standard	ANP	Prod/quality	EF	AZ
Industrial lost time	TPIND	Prod/quality	EF	AZ
Time lost due to harvest	TPFC	Prod/quality	EF	CT
Leveraging RPC	ARPC	Prod/quality	EF	AZ
Agricultural performance	RAGR	Prod/quality	PR	PA CT
Sugar per hectare	AZHA	Prod/quality	PR	PA CT AZ
Empty area	AVAC	Prod/quality	PR	PA
Percentage of electricity delivered	PORC	Prod/quality	PR	AZ COM
Weight profit per weight of sale	PU/PV	Strat/plan	EC	PA CT AZ DER
Potential sugarcane yield	RPC	Prod/quality	PR	PA CT
Dextran in sugar	DEXT	Prod/quality	LIME	AZ COM
Average monthly salary	SALM	Human Des	CH	AZ DER
Worker productivity	PROD	Human Des	CH	AZ DER
Diversification index	DIVER	Innovation	IN	AZ DER COM
Contribution to local development	APORTE	Social responsiveness	EC	PA CT AZ DER COM

The sedimentation graph obtained recommends the analysis of the first five factors (eigenvalue > 1) whose specific and accumulated contributions are shown in Table 2. This selection of factors explains 73.5% of the total variance.

Analyzing the weight of each key performance indicator allows you to rename the factors, focusing on those indicators with the highest coefficient. This exercise is shown in Table 3.

Table 2. Results of the analysis of the first five factors (n = 16) (Statgraphics Centurion XVI)

Factor Number	Eigenvalue	Percentage of Variance	Percentage Accumulated
1	4.09943	25.621	25.621
2	2.91153	18.197	43,819
3	2.35799	14,737	58,556
4	1.36748	8.547	67.103
5	1.01469	6.342	73,445

Table 3. Coefficients of the key performance indicators and name assigned to each factor.

	Factor 1	Factor 2	Factor 3	Factor 4	Factor 5
	Agroindustrial efficiency	Economic/Innovation	Agricultural	Average salary/electricity	Harvest
RIND	0.892	0.151	−0.130	−0.058	−0.146
ANP	0.581	0.354	−0.440	−0.176	−0.120
TPIND	−0.490	−0.506	−0.128	−0.301	−0.012
TPFC	0.406	0.356	0.416	0.224	0.465
ARPC	0.883	0.229	−0.003	−0.159	−0.158
RAGR	−0.392	−0.351	−0.739	0.074	0.267
AZHA	0.115	−0.260	−0.877	0.016	0.157
AVAC	0.064	0.353	0.372	−0.610	0.313
PORC	0.197	−0.140	−0.170	−0.489	−0.393
PU/PV	0.489	−0.651	0.183	0.134	0.291
RPC	0.552	−0.032	−0.314	0.337	−0.035
DEXT	−0.460	−0.412	0.231	−0.413	−0.011
SALM	0.428	−0.121	−0.288	−0.411	0.375
PROD	0.560	−0.666	0.203	−0.077	0.255
DIVER	0.158	−0.697	0.343	0.250	−0.237
APORTE	0.535	−0.670	0.204	−0.080	−0.196

The first factor, with over 25% of the total variance explained, is the factor we call Agroindustrial Efficiency, since the variables with the highest coefficients are of this type. In the linear Eq. (1) that defines its calculation, we can see how industrial yield and RPC utilization have the highest coefficients in the positive direction, while industrial lost time, dextran content, and agricultural yield have the highest coefficients in the opposite direction. Companies with better results in this component achieve greater worker productivity and beneficial average wages, as well as economic profitability. It is noteworthy that a high agroindustrial efficiency factor does not have to be highly diversified to achieve a high factor, since those variables linked to the sugar harvest predominate.

$$\begin{aligned} F1 =& 0.892632 * RIND + 0.581542 * ANP - 0.490994 * TPIND \\ &+ 0.406719 * TPFC + 0.883971 * ARPC - 0.392183 * RAGR + 0.115036 * AZHA \\ &+ 0.064682 * AVAC + 0.197046 * PORC + 0.489484 * PU/PV \\ &+ 0.552397 * RPC - 0.460958 * DEXT + 0.428117 * SALM + 0.56079 * PROD \\ &+ 0.158685 * DIVER + 0.535097 * APORTE \end{aligned} \quad (1)$$

The second factor is the Economic/Innovation factor, which accounts for almost 19% of the variance in an orthogonal sense. In its expression, the economic indicators and the diversification index have the highest coefficients. Figure 2 shows the relationships between factors F1 and F2 on a two-dimensional plane.

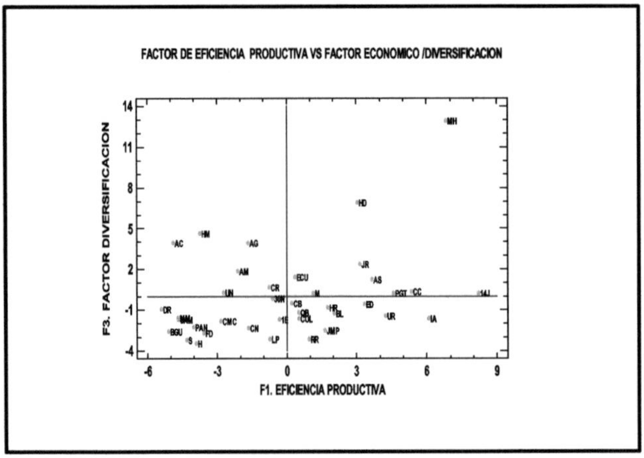

Fig. 2. Two-dimensional plane of F1 vs F2.

Sugar companies with attached distilleries, a high level of diversification, and efficient harvests logically achieve the best economic results (MH and HD). However, other companies with the same start-up conditions (distilleries or other plants) do not achieve better economic results because they are inefficient in harvesting (HM, AG, and AC). A third group (cluster) is identified with the companies discussed above. These companies are very efficient in harvesting but lack the investment to produce higher-value derivatives, and their economic results are positive but very modest (14 J, PGT, and CC).

The third factor integrates the indicators associated with agricultural production with the highest coefficients and explains just under 15% of the total variance. Figure 3 relates this Agricultural factor (F3) to Factor 1, Agroindustrial Efficiency.

Similarly, the companies with the best sugarcane results (fewer empty areas, higher sugarcane yields) occupy privileged positions on the vertical axis, but fall short in terms of productive efficiency (JR, EC, and CR). Some in this cluster (AC and CMC) are even among the worst performers on the horizontal axis.

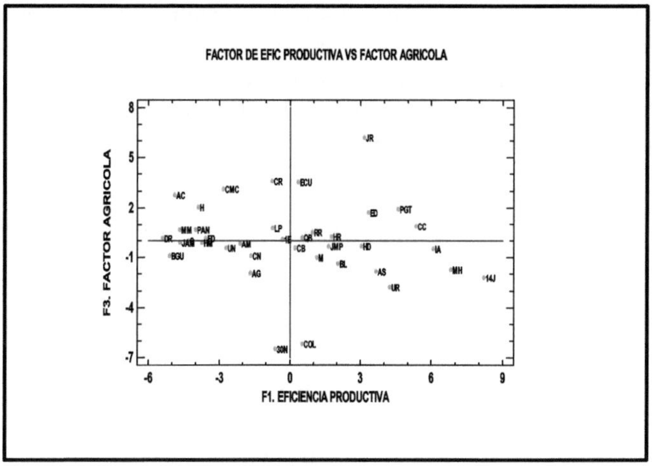

Fig. 3. Two-dimensional plane of F1 vs F3

Two other types of companies can be identified: the MH and 14J types, which are very efficient but do not produce sugarcane, and the 30N and COL types, which do not achieve good results in either of the two areas.

Factors four and five are included in the scree plot, but their specific contribution to variance reduction is not high, falling below 10%. However, they do reveal indicators of interest with high coefficients, such as average wage and electricity (in F4) and time lost during harvesting in the case of F5.

The analysis of the indicators' commonality confirms their appropriate selection, as well as the transversality of those associated with sugarcane production. Sugarcane yield (RAGR) has a value above 0.90, and industrial yield is 0.87. Of all the selected indicators, only the percentage of electricity delivery is below 0.50.

4 Discussion

Equation (2) is proposed for calculating the competitiveness index

$$ICOM = \sum_{I}^{5} qFdi \tag{2}$$

where F is the value of the factor and q is the specific contribution of each factor.

Table 4 contains the value of each factor and the competitiveness index for each of the companies analyzed. Based on the processed data and the assumptions presented, fourteen companies had a positive result on the indicator. The most competitive sugar companies during the period analyzed were MH, JR, and HD, with economic stability and results that set them apart from the rest.

Table 4. Value of the factors and the competitiveness index in each company.

	Agroindustrial Efficiency	Economic/Innovation	Agricultural	Average salary/electricity delivery	Harvest	ICOM
MH	6.832	12.917	−1.704	2.669	1.307	3.50
JR	3.149	2.420	6.242	−0.483	0.959	1.77
HD	3.041	6.921	−0.323	−1.571	−0.838	1.41
CR	−0.749	0.708	3.630	0.586	5.575	1.41
HM	−3.742	4.627	−0.092	1.214	0.067	1.35
ECU	0.346	1,463	3.575	1,210	0.991	1.29
AC	−4.892	3.942	2.748	−2.421	−0.025	1.15
DC	5.338	0.394	0.878	2.329	1.055	0.69
AM	−2.114	1.904	−0.164	−0.155	0.103	0.44
AG	−1.685	3.947	−1,950	−2.804	1,539	0.37
PGT	4.591	0.298	1.919	0.091	−1.265	0.33
CMC	−2.822	−1.795	3.121	−0.292	1954	0.23
14J	8.213	0.252	−2.182	3.278	0.416	0.19
ED	3.342	−0.545	1,749	−1.175	0.884	0.08
AS	3.652	1.247	−1.844	−0.283	−0.442	−0.10
HR	1,766	−0.798	0.269	0.032	0.307	−0.12
QB	0.513	−1.167	0.219	0.613	0.389	−0.14
M	1,150	0.294	−0.992	−0.652	0.471	−0.16
PAN	−4.017	−2.257	0.686	1,538	0.272	−0.20
A	−2.719	0.350	−0.433	−1.378	−0.456	−0.23
AI	6.091	−1.635	−0.484	2.259	−1.083	−0.27
DR	−5.398	−0.937	0.162	−0.570	0.093	−0.29
JMP	1,657	−2.474	−0.298	3.005	−0.840	−0.32
CB	0.225	−0.475	−0.406	−0.603	−0.694	−0.34
MM	−4.680	−1.579	0.666	−1.124	1.128	−0.35
1E	−0.303	−1.672	0.103	0.252	−0.367	−0.40
BL	2.039	−1.226	−1.364	0.201	0.139	−0.52
JAM	−4.665	−1.653	−0.087	−1.273	−0.361	−0.66

(continued)

Table 4. (*continued*)

	Agroindustrial Efficiency	Economic/Innovation	Agricultural	Average salary/electricity delivery	Harvest	ICOM
UR	4.230	−1.424	−2.761	1,270	−1.612	−0.82
BGU	−5.087	−2.520	−0.904	0.244	−0.794	−0.84
RR	0.953	−3.094	0.535	−1.027	−0.118	−0.86
FD	−3.579	−2.693	0.140	−1.549	−0.001	−0.89
LP	−0.726	−3.124	0.779	−1.525	−0.450	−0.92
S	−4.312	−3.158	0.026	−1.328	−0.220	−1.02
CN	−1.651	−2.320	−0.898	−1.653	−1.221	−1.11
H	−3.886	−3.424	2.046	−2.443	−3.757	−1.19
30N	−0.621	−0.081	−6.448	0.851	−1.434	−1.19
COL	0.514	−1.631	−6.158	2.666	−1.671	−1.29

The selected scenario (2020–2024) could not be worse. There will be an absolute decrease in the area covered by sugarcane, a cessation of exports, and a sharp deterioration in the working and living conditions of workers in the sector. However, there are also positive elements, such as a different perspective on local linkages, agricultural diversification with food production, and, in general, the need for innovation and the introduction of new scientific results.

The highest ICOM competitiveness index score obtained with the data matrix obtained during the period analyzed is obtained by companies MH, JR, and HD, making them the most competitive during the period analyzed. The ICOM score for an "ideal" company is obtained by using the highest scores obtained for each of the factors, resulting in an index of 5.41. MH's score would therefore be just 65%, and it would have to significantly improve its agricultural and harvesting factors to occupy the top position.

The points earned by JR for its strong sugarcane performance (+6) contrast with those lost by MH (−1.7) and 14J (−2.18) for the same agricultural factor (F3). This poor performance in sugarcane production by many companies affects their overall competitiveness indicator. 14J, for example, one of the sugarcane companies that has stood out for its harvest efficiency in recent years, with a score of + 8 in the first factor, falls to 13th place in the ranking, although it still has a positive ICOM.

All companies with an attached distillery, producers of alcohol and rum, and others that sell molasses even if they don't produce sugar, consistently receive a positive competitiveness rating, which reaffirms the relationship between the circular economy and diversification and the company's overall results.

5 Conclusions

1. The new ICOM competitiveness index allows for a comprehensive assessment of sugarcane value chain management in sugar companies.

2. The direct relationship between diversification and economic and social results is confirmed. Investing in the construction of new derivative plants or incorporating new products in sugar companies that do not do so certainly has a positive impact on the organization's overall results and competitiveness.
3. The agricultural sector plays a fundamental role in increasing the competitiveness of our sugar companies. The positioning of the organizations involved depends primarily on the coverage of the area and increased sugarcane yields.
4. The results of these five factors and ICOM should be compared to previous stages that had better results in sugarcane production, the exploitation of productive capacities, and greater exports. This will surely result in a higher competitiveness indicator and could serve as a reference for complementary evaluations.

References

1. Aguilar, N.: Value chain of the diversification of the sugarcane agroindustry in Mexico. Agroproductividad **10**(11), 21–28 (2017)
2. Porter, M.: From Competitive Advantage to Business Strategy. Harvard Business Review. Annotated translation published in the Cuadernos de Finanzas (Financial Notebooks) of the Argentine Society of Business Administration (SADAF), No. 20, 1992 (1987). Available at http://www.sadaf.com.ar/index.php?act=showCuadernos
3. Sarmiento Reyes, Y.R., Delgado Fernández, M.: Measuring business competitiveness for territorial development. COODES **8**(3), 409–424 (2022)
4. Rein, P.: Sustainable production of raw and refined dog sugar. In: Presented at SIT Conference, Montreal, p. 11 (2011). Available at https://bonsucro.com/assets/rein_paper.pdf
5. Ramírez Molina, R. Ampudia Sjogreen, D.: Business competitiveness factors in the commercial sector. Electron. J. Sci. Technol. Maracaibo Univ. Inst. Technol., **4**(1) (2018)
6. Casanova, E., Almazán, O. Considerations on projections for a highly competitive sugar agroindustry. Annals Cuban Acad. Sci., **5**(2) (2015). Available at: http://www.revistaccuba.cu/index.php/revacc/issue/view/11/showToc
7. Godet, M.: The Toolbox of Strategic Foresight. LIPSOR, Paris (2009)
8. Mojica, F.: Future Studies: Linearity vs. Plurality. Technical Secretariat of the National Science and Technology Organizations of the Andrés Bello Convention Countries (ONCYT/CAB), Bogota, Colombia (2004). Available at http://www.cab.int.co/www/cab3/downloads/prospectiva−3.pdf
9. Garrido Bermúdez, E., Mena Rodríguez, H.Y., Zuluaga Arango, J.M., Pérez Quintero, F.E.: Process for validating a research instrument through factor analysis. UNACIENCIA **16**(30), 61–73 (2023). https://doi.org/10.35997/unaciencia.v16i30.724
10. Caballero, A.: What is the true agro−productive potential of sugarcane? Icidca J. Sugarcane Deriv. **57**(3), 28–35 (2023)
11. Caballero, A., Sabadi, R.: The sugarcane value chain: its strengths and weaknesses in present−day Cuba. In: Conference Workshop on the Sugar Value Chain, Agrarian University of Havana (2024)

Author Index

A. Santos Fuentefria et al. (Eds.): JIFREEEMA 2025, LNNS 1933, pp. 209–210, 2026.
https://doi.org/10.1007/978-3-032-23028-7

MIX
Papier aus verantwortungsvollen Quellen
Paper from responsible sources
FSC® C105338

If you have any concerns about our products,
you can contact us on
ProductSafety@springernature.com

In case Publisher is established outside the EU,
the EU authorized representative is:
Springer Nature Customer Service Center GmbH
Europaplatz 3, 69115 Heidelberg, Germany

Printed by Libri Plureos GmbH
in Hamburg, Germany